Foundations of Electronics Laboratory Projects, 5e

to accompany

Foundations of Electronics: 5e (Electron Flow Version)
Foundations of Electronics: Circuits and Devices, 5e
(Electron Flow Version)
and
Foundations of Electronics: Circuits and Devices, 2e
(Conventional Flow version)

Russell L. Meade

Australia • Brazil • Japan • Korea • Mexico • Singapore • Spain • United Kingdom • United States

Foundations of Electronics Laboratory Projects, Fifth Edition

Russell L. Meade

Vice President, Technology and Trades ABU: Dave Garza

Director of Learning Solutions: Sandy Clark

Senior Acquisitions Editor: Stephen Helba

Senior Product Manager: Michelle Ruelos Cannistraci

Channel Manager: Dennie Williams

Marketing Coordinator: Stacey Wiktorek

Production Director: Mary Ellen Black

Senior Production Manager: Larry Main

Marketing Director: Deborah S. Yarnell

Production Editor: Benj Gleeksman

Art/Design Coordinator: Francis Hogan

Senior Editorial Assistant: Dawn Daugherty

© 2007 Delmar, Cengage Learning

ALL RIGHTS RESERVED. No part of this work covered by the copyright herein may be reproduced, transmitted, stored or used in any form or by any means graphic, electronic, or mechanical, including but not limited to photocopying, recording, scanning, digitizing, taping, Web distribution, information networks, or information storage and retrieval systems, except as permitted under Section 107 or 108 of the 1976 United States Copyright Act, without the prior written permission of the publisher.

> For product information and technology assistance, contact us at
> **Cengage Learning Customer & Sales Support, 1-800-354-9706**
> For permission to use material from this text or product,
> submit all requests online at **www.cengage.com/permissions**
> Further permissions questions can be emailed to
> **permissionrequest@cengage.com**

Library of Congress Control Number: 2005046743

ISBN-13: 978-1-4180-4183-0

ISBN-10: 1-4180-4183-1

Delmar
Executive Woods
5 Maxwell Drive
Clifton Park, NY 12065
USA

Cengage Learning is a leading provider of customized learning solutions with office locations around the globe, including Singapore, the United Kingdom, Australia, Mexico, Brazil, and Japan. Locate your local office at **international.cengage.com/region**

Cengage Learning products are represented in Canada by Nelson Education, Ltd.

For your course and learning solutions, visit **delmar.cengage.com**

Visit our corporate website at **www.cengage.com**

Notice to the Reader
Publisher does not warrant or guarantee any of the products described herein or perform any independent analysis in connection with any of the product information contained herein. Publisher does not assume, and expressly disclaims, any obligation to obtain and include information other than that provided to it by the manufacturer. The reader is expressly warned to consider and adopt all safety precautions that might be indicated by the activities described herein and to avoid all potential hazards. By following the instructions contained herein, the reader willingly assumes all risks in connection with such instructions. The publisher makes no representations or warranties of any kind, including but not limited to, the warranties of fitness for particular purpose or merchantability, nor are any such representations implied with respect to the material set forth herein, and the publisher takes no responsibility with respect to such material. The publisher shall not be liable for any special, consequential, or exemplary damages resulting, in whole or part, from the readers' use of, or reliance upon, this material.

Printed in the United States of America
7 8 9 10   13 12 11

CONTENTS

Preface vii
Introduction ix
Instructions for Troubleshooting Exercises xiii
Breadboarding Circuits Information xv

PART 1: USE AND CARE OF METERS — 1

PROJECT 1 Voltmeters 3
PROJECT 2 Ammeters 5
PROJECT 3 Ohmmeters 9

PART 2: OHM'S LAW — 13

PROJECT 4 Resistor Color Code Review and Practice 15
PROJECT 5 Relationship of I and V with R Constant 19
PROJECT 6 Relationship of I and R with V Constant 21
PROJECT 7 Relationship of Power to V with R Constant 25
PROJECT 8 Relationship of Power to I with R Constant 29
                Story Behind the Numbers: Ohm's Law 31

PART 3: SERIES CIRCUITS — 37

PROJECT 9   Total Resistance in Series Circuits 39
PROJECT 10 Current in Series Circuits 41
PROJECT 11 Voltage Distribution in Series Circuits 43
PROJECT 12 Power Distribution in Series Circuits 47
PROJECT 13 Effects of an Open in Series Circuits 51
PROJECT 14 Effects of a Short in Series Circuits 55
                Story Behind the Numbers: Series Circuits 57

PART 4: PARALLEL CIRCUITS — 65

PROJECT 15 Equivalent Resistance in Parallel Circuits 67
PROJECT 16 Current in Parallel Circuits 73
PROJECT 17 Voltage in Parallel Circuits 75
PROJECT 18 Power Distribution in Parallel Circuits 77
PROJECT 19 Effects of an Open in Parallel Circuits 79
PROJECT 20 Effects of a Short in Parallel Circuits 83
                Story Behind the Numbers: Parallel Circuits 85

PART 5: SERIES-PARALLEL CIRCUITS — 93

PROJECT 21 Total Resistance in Series-Parallel Circuits 95
PROJECT 22 Current in Series-Parallel Circuits 99

PROJECT 23 Voltage Distribution in Series-Parallel Circuits 103
PROJECT 24 Power Distribution in Series-Parallel Circuits 105
PROJECT 25 Effects of an Open in Series-Parallel Circuits 107
PROJECT 26 Effects of a Short in Series-Parallel Circuits 111
Story Behind the Numbers: Series-Parallel Circuits 113

PART 6: BASIC NETWORK THEOREMS 123

PROJECT 27 Thevenin's Theorem 125
PROJECT 28 Norton's Theorem 129
PROJECT 29 Maximum Power Transfer Theorem 133
Story Behind the Numbers: Basic Network Theorems 135

PART 7: NETWORK ANALYSIS TECHNIQUES 143

PROJECT 30 Loop/Mesh Analysis 145
PROJECT 31 Nodal Analysis 149
PROJECT 32 Wye-Delta and Delta-Wye Conversions 153
Story Behind the Numbers: Network Analysis Techniques 157

PART 8: THE OSCILLOSCOPE 163

PROJECT 33 Basic Operation: Familiarization 165
PROJECT 34 Basic Operation: Controls Manipulation 169
PROJECT 35 Basic Operation: Vertical Controls and DC V 171
PROJECT 36 Basic Operation: Observing Various Waveforms 173
PROJECT 37 Voltage Measurements 175
PROJECT 38 Phase Comparisons 177
PROJECT 39 Determining Frequency 179

PART 9: INDUCTANCE 183

PROJECT 40 Total Inductance in Series and Parallel 185
Story Behind the Numbers: Inductance 187

PART 10: INDUCTIVE REACTANCE IN AC 193

PROJECT 41 Induced Voltage 195
PROJECT 42 Relation of X_L to L and Frequency 197
Story Behind the Numbers: Inductive Reactance in AC 203

PART 11: *RL* CIRCUITS IN AC 209

PROJECT 43 V, I, R, Z, and θ Relationships in a Series *RL* Circuit 211
PROJECT 44 V, I, R, Z, and θ Relationships in a Parallel *RL* Circuit 215
Story Behind the Numbers: *RL* (Series) Circuits in AC 219
Story Behind the Numbers: *RL* (Parallel) Circuits in AC 223

PART 12: BASIC TRANSFORMER CHARACTERISTICS — 229

PROJECT 45 Turns, Voltage, and Current Ratios 231
PROJECT 46 Turns Ratios versus Impedance Ratios 233
　　　　Story Behind the Numbers: Basic Transformer Characteristics 235

PART 13: CAPACITANCE (DC CHARACTERISTICS) — 241

PROJECT 47 Charge and Discharge Action and RC Time 243
PROJECT 48 Total Capacitance in Series and Parallel 245
　　　　Story Behind the Numbers: Capacitance (DC Characteristics) 247

PART 14: CAPACITIVE REACTANCE IN AC — 253

PROJECT 49 Capacitance Opposing a Change in Voltage 255
PROJECT 50 X_C Related to Capacitance and Frequency 257
PROJECT 51 The X_C Formula 261
　　　　Story Behind the Numbers: Capacitive Reactance in AC 263

PART 15: *RC* CIRCUITS IN AC — 269

PROJECT 52 V, I, R, Z, and θ Relationships in a Series RC Circuit 271
PROJECT 53 V, I, R, Z, and θ Relationships in a Parallel RC Circuit 275
　　　　Story Behind the Numbers: RC Circuits in AC 279

PART 16: SERIES RESONANCE — 287

PROJECT 54 X_L and X_C Relationships to Frequency 289
PROJECT 55 V, I, R, Z, and θ Relationships When $X_L = X_C$ 291
PROJECT 56 Q and Voltage in a Series Resonant Circuit 295
PROJECT 57 Bandwidth Related to Q 299
　　　　Story Behind the Numbers: Series Resonance 303

PART 17: PARALLEL RESONANCE — 311

PROJECT 58 V, I, R, Z, and θ Relationships when $X_L = X_C$ 313
PROJECT 59 Q and Impedance in a Parallel Resonant Circuit 317
PROJECT 60 Bandwidth Related to Q 319
　　　　Story Behind the Numbers: Parallel Resonance 321

PART 18: THE SEMICONDUCTOR DIODE — 329

PROJECT 61 Forward and Reverse Bias, and I vs. V 331
PROJECT 62 Diode Clipper Circuits 335

PART 19: SPECIAL-PURPOSE DIODES — 343

PROJECT 63 Zener Diodes 345
PROJECT 64 Light-Emitting Diodes 349

PART 20: POWER SUPPLIES — 353

- **PROJECT 65** Half-Wave Rectifier 355
- **PROJECT 66** Bridge Rectifier 359
- **PROJECT 67** Voltage Multiplier 363
    - Story Behind the Numbers: Characteristics of a 7805 Voltage Regulator 365

PART 21: BJT CHARACTERISTICS — 373

- **PROJECT 68** BJT Biasing 375
    - Story Behind the Numbers: BJT Transistor Characteristics 379

PART 22: BJT AMPLIFIER CONFIGURATIONS — 389

- **PROJECT 69** Common-Emitter Amplifier 391
- **PROJECT 70** Common-Collector Amplifier 395

PART 23: BJT AMPLIFIER CLASSES OF OPERATION — 401

- **PROJECT 71** BJT Class A Amplifier 403
- **PROJECT 72** BJT Class B Amplifier 407
- **PROJECT 73** BJT Class C Amplifier 411

PART 24: JFET CHARACTERISTICS AND AMPLIFIERS — 417

- **PROJECT 74** Common-Source JFET Amplifier 419
- **PROJECT 75** Power FETs 423
    - Story Behind the Numbers: JFET Transistor Characteristics 431

PART 25: OPERATIONAL AMPLIFIERS — 441

- **PROJECT 76** Inverting Op-Amp Circuit 443
- **PROJECT 77** Noninverting Op-Amp Circuit 447
- **PROJECT 78** Op-Amp Schmitt-Trigger Circuit 451

PART 26: OSCILLATORS AND MULTIVIBRATORS — 457

- **PROJECT 79** Wien-Bridge Oscillator Circuit 459
- **PROJECT 80** 555 Astable Multivibrator 463

PART 27: SCR OPERATION — 467

- **PROJECT 81** SCRs in DC Circuits 469
- **PROJECT 82** SCRs in AC Circuits 473

PART 28: FIBER-OPTIC SYSTEM CHARACTERISTICS — 479

- Story Behind the Numbers: Fiber-Optic System Characteristics 479

- **APPENDIX A** Parts Suppliers Listing 487
- **APPENDIX B** How to Create a Graph or Chart Using Excel 489

PREFACE

Purpose

- To confirm and reinforce theory concepts
- To provide hands-on experience in connecting circuits from schematics, making measurements, organizing data, creating graphs, and analyzing observations
- To improve critical thinking skills

Features

1. **Companion to Fundamentals of Electronics texts.**

   Projects are conveniently and directly correlated with the companion textbooks, *Foundations of Electronics* (Electron Flow Version, 5e), *Foundations of Electronics: Circuits and Devices* (Electron Flow Version, 5e), and *Foundations of Electronics: Circuits and Devices* (Conventional Flow Version, 2e). These projects can also be used effectively in conjunction with any introductory dc/ac text.

2. **Flexible format and approach.**

   The lab manual offers two different format styles for many projects.

   - **Workbook-style format.** Most projects retain the workbook style to promote understanding of the points made by taking measurements and performing calculations and then drawing conclusions from them at optimum times during the project.
   - **Story Behind the Numbers format.** The lab manual includes projects, titled *Story Behind the Numbers*. These projects provide appropriate practice in connecting circuits from schematics, making measurements and calculations, collecting data, creating graphs, answering analysis questions, and writing technical lab reports.

3. **An optional choice of lower-frequency or higher-frequency operation is provided in a number of projects involving inductors.**

   This allows instructors to select their preference of the frequency of operation, whether the constraints are component or equipment availability, or instructor preferences for learning purposes.

Advice to Students

- Read "How to Use This Lab Manual" in the Introduction, which follows this Preface.
- Do not be concerned if you do not know the exact word called for in the blank spaces of the Conclusion sections. The important thing is that you understand the concept being highlighted and can give a word or words that convey your understanding.
- Your rewards will be directly related to your effort.

We wish you great success in your experiences and trust that this guide will be both interesting and meaningful as you prepare to enter the world of electronics technology.

INTRODUCTION

How to Use This Lab Manual

The lab manual is divided into 28 parts. Each part consists of a group of projects. There are a total of 82 workbook-style projects and 20 new *Story Behind the Numbers* projects.

Part Openers

Each part begins with

- **Objectives**, which list knowledge and skills to be acquired while completing projects, and
- **Project/Topic Correlation Information**, which lists the corresponding textbook chapters and appropriate topics within the chapters that are related to each project.

Workbook-Style Projects

These projects are designed to help students develop and improve their abilities to

- Follow instructions carefully.
- Make accurate measurements and calculations.
- Draw logical conclusions from their observations and calculations.

Each *workbook-style* project contains some or all of the following elements:

- **Graphics and/or Schematics** illustrating how equipment and components should be connected and how to set up the circuit.
- **Project Purposes** describing objectives of the lab experiment.
- **Parts Needed** identifying equipment and components needed.
- **Safety Hints, Cautions, and Notes** highlighting warnings and practical lab reminders.
- **Procedures** outlining the steps involved, such as connecting, applying, measuring, calculating, and so on.
- **Observations** providing space where observations, measurements, calculations, and/or results can be recorded.
- **Conclusions** requiring students to analyze key concepts that relate to their observations.

"Story Behind the Numbers" Projects

These projects are designed to help students develop and improve their abilities to

- Follow instructions carefully.
- Make accurate measurements and calculations.
- Create tables and graphs for presenting collected data.
- Analyze technical data appropriately.
- Write technical reports and communicate clearly and accurately.

Each *Story Behind the Numbers* project contains the following elements:

- **Procedures** defining circuits to be connected, measurements and calculations to be made, and tasks to perform, such as collecting and displaying data.
- **Analysis Questions** requiring in-depth thinking and analyses regarding the student's measurements, calculations, and graphic data presentations.
- **A Technical Lab Report** providing an opportunity for students to practice and sharpen their technical writing and communications skills.

The following example illustrates a basic sample of how a Technical Lab Report might look:

Sample Technical Lab Report

1. This report relates to the project covering the series circuit configuration. The key parameters observed and analyzed for this project included the circuit current, voltage drops and voltage distribution characteristics, and power dissipations and distribution characteristics.
2. Observations, measurements, and calculations performed using this project's circuit highlighted the relationships between circuit voltage, resistance and current, voltage distribution related to component resistance values, and power dissipations and distribution related to component resistance values.
3. The Analysis Questions pointed out that the basic characteristic of this type of circuit that makes it unique from other circuit configurations was related to the fact that there is only *one path* for current; thus, current through all components is "one and the same" current. This is highlighted by the fact that the voltage drops and power dissipations for each component are computed by using the same current value.
4. Some practical examples of how knowledge gained from analyzing this type of circuit might be used elsewhere include troubleshooting house wiring between light switches and lights, and understanding numerous automotive circuits using series circuits. In addition, the "half-split" troubleshooting rule is very applicable to this type of circuit and any other system having "in-line" components or elements.
5. My appraisal regarding the positive aspects of this project is that the instructions were clear and concise. The measurements, calculations, and analyses requirements clearly focused my attention on the key parameters for this type of circuit.

Specific Notations for the User

When using the "higher-frequency" circuit options, pay attention to the cautions regarding DMM measuring limitations (frequency limits). If using a DMM, be sure it is rated to properly read voltages at the frequencies required in the project.

If a scope is to be used, observe the critical ground connection precautions necessary to prevent shorting out circuit components or portions of the circuit.

Equipment

The general types of equipment required to perform the project experiments include:

1. Component interconnection system (CIS) (proto boards or similar connection matrices)
2. Variable voltage dc power supply (VVPS)
3. DMM
4. Function generator
5. Oscilloscope (dual-trace, if possible)
6. Digital frequency counter (optional)

Component Parts

The component parts required in order to perform all the projects in this manual include:

For Projects 1–60

Resistors

Resistor decade box
Kit of 10 various color-coded values
½ W resistors:
(Note: In most cases ¼ W resistors are sufficient.)
100 Ω (2)
270 Ω
780 Ω
820 Ω
1 kΩ (4)
3.3 kΩ
4.7 kΩ
5.6 kΩ
10 kΩ (4)
12 kΩ
18 kΩ
27 kΩ
47 kΩ
100 kΩ (2)
1 MΩ
10 MΩ
Potentiometer:
100 kΩ

Inductors

One inductor with very high L/R ratio
1.5 H, 95 Ω (dc) (2) or appropriate equivalents
100 mH (2)

Capacitors

0.1 µF (2)
1.0 µF (2)

Transformers

12.6 V (w/ center-tapped secondary)
Audio output transformer
Isolation transformer

Miscellaneous Items

Dry cell, 1.5 V
Speaker, 4 Ω
Meter:
  0–1 mA (dc) panel meter (optional)

For (Devices) Projects 61–82

Resistors

47 Ω
100 Ω
220 Ω
270 Ω
330 Ω
470 Ω
1 kΩ (2)
1.5 kΩ (2)
2.0 kΩ (2)
2.7 kΩ
4.7 kΩ
6.8 kΩ
10 kΩ (4)
12 kΩ
22 kΩ
27 kΩ
47 kΩ
100 kΩ (2)
270 kΩ
470 kΩ
1 MΩ
1 kΩ linear potentiometer
1 kΩ trim potentiometer
10 kΩ trim potentiometer
Carbon, 1/4 or 1/2 W
10 Ω (5 W)
20 Ω (2 W)
330 Ω (1/2 W)

Transistors/Semiconductors

NPN, silicon, 2N3904 or equivalent
N-channel JFET, 2N5458 or equivalent
Op-Amp, 741
555 Timer IC
SCR, 2N5060, or equivalent
N-channel power FET IRF510
P-channel power FET IRF9530

Diodes

Silicon diode: 1-ampere rating (1N4002 or equivalent) (4)

LED: gallium arsenide red, 20 mA
Zener diode: 5.1 V, 1 W (1N4733 or equivalent)

Capacitors

0.01 µF (2)
0.1 µF (2)
0.33 µF
1.0 µF (2)
10 µF (2)

Miscellaneous Items

1.5-V cell
6-V incandescent lamp (#47)
±15-V DC power supply
6.3-V AC source
DMM/VOM (2)
DC variable voltage power supply
Function/signal generator
Dual-trace oscilloscope
CIS
12-V DC motor (Radio Shack 273-256)

Acknowledgments

A note of thanks to the following reviewers:

Don Barrett, DeVry University, Irving, TX
Gary Cardinale, DeVry University, New Brunswick, NJ
Phil Golden, DeVry University, Kansas City, MO
Marcus Rasco, DeVry University, Irving, TX

INSTRUCTIONS FOR OPTIONAL TROUBLESHOOTING EXERCISES

The Optional Troubleshooting Exercises are designed to provide hands-on troubleshooting experiences that allow students to apply the SIMPLER troubleshooting process to actual circuits, set up by their instructors.

Note that each of the "Student Log for Optional Troubleshooting Exercise" sheets is set up to lead you through the **SIMPLER** troubleshooting sequence. That is:

- In the first blank on the log sheet, you must document the starting point **S**ymptom data.
- In the next blank, you must identify the **I**nitial "area of uncertainty" or the area of the circuit you would bracket for troubleshooting efforts.
- The third blank requires you to log the decision you have made regarding your first test (i.e., your response to the **M**ake a decision about what type of test and where to make it step). At this point, you should **P**erform the test and observe the results of the test.
- The next step is to use the information you have collected so far to help you log data that **L**ocate and enable you to define the new (hopefully smaller) area to be bracketed for continued troubleshooting.
- The next blank on the log sheet requires you to **E**xamine your available data and determine the next test you want to make. Log the type and location of the test. Now you make the test and observe results that enable you to **R**epeat the analysis and testing steps (as indicated on the log sheet) until you find what you believe to be the problem with the circuit.

NOTE The instructor's part in this effort is to supply the student with pre-wired circuits having specific problems in them, as indicated in the Instructor's Guide. Problem circuits should be supplied to the student on a one-circuit-problem-at-a-time basis.

BREADBOARDING CIRCUITS INFORMATION

To "breadboard" a circuit requires properly interconnecting components, jumpers, and electrical devices in such a way as to conform to the connections of a specified circuit. This breadboarded circuit then allows for practical measurements and analysis of how the circuit specified in a schematic or other diagram actually operates.

Sample Circuit Board Matrix

Although there are a number of types of circuit connection devices available, a typical type is shown below. The board is made of plastic with a matrix of holes, into which wires and component leads are pushed to make appropriate connections. Refer to the figure below as you study the following points regarding this type of circuit board.

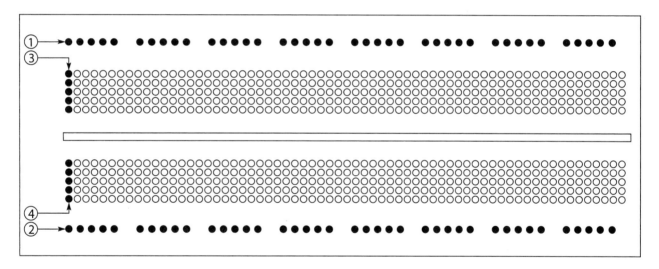

- Each "hole" on the board contains a metallic spring contact. This means that an electrical wire, or component lead pushed down into the hole, is making electrical contact with that hole's spring contact.
- The circuit board provides automatic "interconnection" between select holes on the circuit board via metallic "bus" connections made underneath the holes. The matrix of holes is internally interconnected so that

    1. The eight horizontal "groups of five holes" along the top are connected in common.
    2. The eight horizontal "groups of five holes" along the bottom are connected in common.
    3. Each vertical column of five holes (above the center separator groove) are interconnected; however, each of these sets of five holes is isolated from

xv

the horizontally adjacent set of five holes, as well as being isolated from all the holes below the board's center separator groove.

4. The five holes in each vertical column (below the center separator groove) are interconnected; however, each set of five holes is isolated from the horizontally adjacent set of five holes, as well as being isolated from all the holes above the board's center separator groove.

General Comments

Because of the typical circuit board layout (and the way buses are made by connected-in-common spring contacts, or built-in interconnections), the following techniques are quite commonly used:

- **When connecting power-supply or signal source leads to the circuit board matrix:**

    1. Connect one source lead to one of the top horizontal row (common bus) holes.
    2. Connect the other source lead to a hole in the bottom horizontal row (common bus).
    3. Insert the jumper lead from each source connection row to the appropriate points in the experimental circuit constructed on the vertical columns portion of the circuit board.

    NOTE ▶ If using alligator clip leads from the source, connections can be made directly to the experimental circuit's component leads, as appropriate, without having to connect the source's leads to the outside rows and then to jumpers to connect to the experimental circuit source input points.

- **When connecting component leads:** Plug one lead from a component into a vertical column hole and the other lead from that particular component into another vertical column hole (horizontally spaced as convenient for the size of the component).

- **When making a connection from one component to the next in a circuit:** Connect one lead from the second component to a lead from the first component. To do so, insert one of the leads from the second component into an adjacent hole in the *same* (set of five) vertical column group as the first component's lead is connected.

Sample Circuit Setups

The following sample circuit setups (interconnections) are taken from illustrations used in the *Foundations of Electronics* text used to portray several of the "Troubleshooting Challenge" circuits. Be aware that the circuit boards illustrated in these portrayals use that same "bussing" technique for interconnections we have been discussing. (You may notice that the circuit boards used in these illustrations have two horizontal rows of holes at the top and the bottom, rather than just one. The interconnection concepts, however, are the same as we have been discussing.)

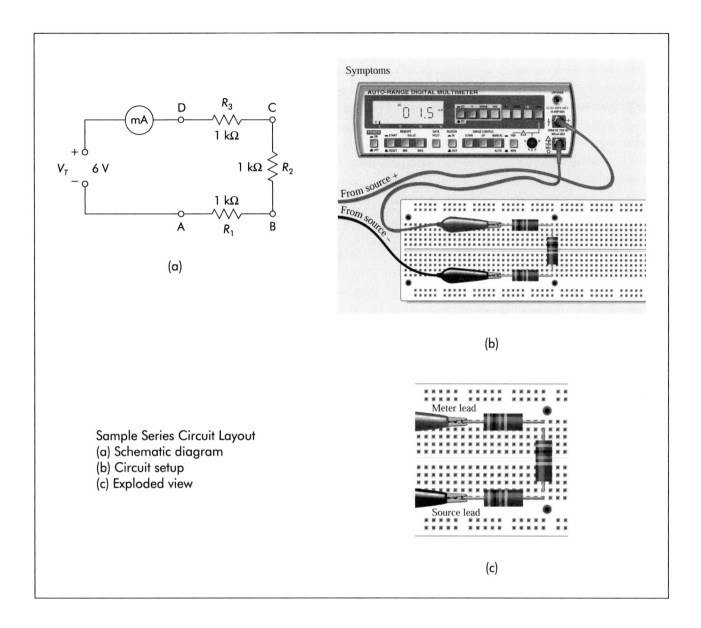

Sample Series Circuit Layout
(a) Schematic diagram
(b) Circuit setup
(c) Exploded view

NOTE ▶ When using the alligator clip leads from the source and meter, as used in the illustration here, we are *not* connecting the source and meter leads to the "outside" horizontal bus rows and then using jumpers from those rows to connect to the circuit points. If alligator clip leads were not used, it would be typical to connect the source leads to the outside horizontal bus points and then use a jumper wire to the experimental circuit points, as appropriate.

BREADBOARDING CIRCUITS INFORMATION

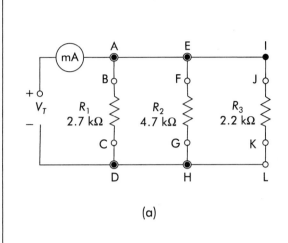

(a)

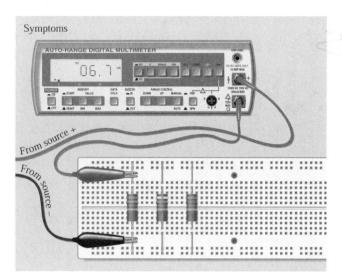

(b)

Sample Parallel Circuit Layout
(a) Schematic diagram
(b) Circuit setup
(c) Exploded view

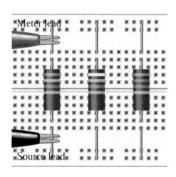

(c)

USE AND CARE OF METERS

PART 1

Objectives

You will become familiar with the basic operation of meters in measuring dc voltage and current. You will also study the basic use of an ohmmeter and learn the normal operating procedures used to protect meter circuits when making measurements.

In completing these projects, you will observe meters, make measurements, draw conclusions, and be able to answer questions about the following items related to the use and care of meters:

- Use of mode/function selector switch(es) (as appropriate)
- Use of range switch(es) (where appropriate)
- Use of test leads
- Proper polarity (where appropriate)
- Proper connection to make voltage measurements
- Proper connection to make current measurements
- Checking continuity with an ohmmeter
- Measuring resistance with an ohmmeter

Project/Topic Correlation Information

PROJECT	TEXT CHAPTER	SECTION	RELATED TEXT TOPIC(S)
1 Voltmeters	2	2-6	Making a Voltage Measurement with a Voltmeter
2 Ammeters	2	2-6	Making a Current Measurement with an Ammeter
3 Ohmmeters	2	2-6	Making a Resistance Measurement with an Ohmmeter

Use and Care of Meters
Voltmeters

Name: James Quinn   Date: 9-11-14

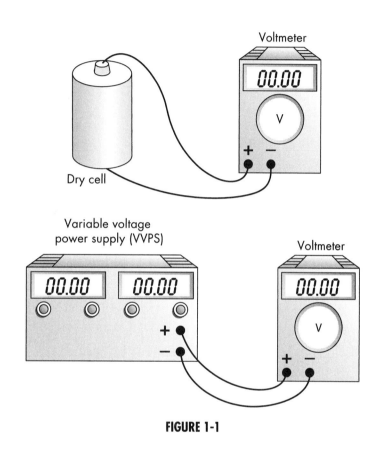

FIGURE 1-1

PROJECT PURPOSE   To learn how to properly and safely measure dc voltages and to acquire practice in connecting simple circuits from diagrams and in adjusting variable dc source voltage(s) for a desired level of output.

PARTS NEEDED   ☐ DMM   ☐ VVPS (dc)
   ☐ 1.5-V dry cell

SAFETY HINTS   Be sure power is off on the VVPS when connecting the meter.

In this project you will begin learning some of the standard procedures for safely using meters to make measurements. To reinforce your thinking, we will take this opportunity to stress key points of meter care in brief form. Think of and properly apply these key factors *every time you use a meter*!

PART 1: Use and Care of Meters

🛑 Safety Hints

1. Use the proper METER MODE/FUNCTION (dc or ac, volts, amperes/mA, or ohms).
2. On nonautoranging meters, be sure the RANGE is high enough for what you will measure.

    NOTE ➤ If not sure, START with the HIGHEST range switch position and work down until the reading is easiest to interpret.

3. Be sure to OBSERVE POLARITY when measuring dc.
4. When MEASURING VOLTAGE, be sure meter is connected IN PARALLEL with the two points having the potential difference to be measured.
5. Use PERSONAL SAFETY cautions! (Power off when connecting test leads or holding only one lead with other hand in pocket, and so on.)

PROCEDURE

1. Use the precautions listed above and use a DMM (multimeter) to measure the voltage of a dry cell (e.g., flashlight, battery, and so on).

    ⚠ OBSERVATION   Cell voltage measures __1.5__ V.

    ⚠ CONCLUSION   Mode/function used was (dc, ac) __dc__ volts. The red test lead was connected to the (+, −) __+__ terminal of the cell. The black test lead was connected to the (+, −) __−__ terminal of the cell. If not autoranging, the range selector switch was in the __V__ voltage range position.

2. If a variable-voltage power supply is available, use the voltmeter to monitor the power supply's output voltage terminals and carefully adjust the power supply to 5-V, 10-V, and 15-V output settings, in that order. Have the instructor check your setting each time.

    ⚠ OBSERVATION   5-V setting OK.   ➤ Instructor initial: _____

    10-V setting OK.   ➤ Instructor initial: _____

    15-V setting OK.   ➤ Instructor initial: _____

    ⚠ CONCLUSION   What mode/function was the meter set in? __dc volts__. The red test lead was connected to the (+, −) __+__ output terminal of the power supply. The black test lead was connected to the (+, −) __−__ output terminal of the power supply. What meter voltage range setting was used? __4V__.

Use and Care of Meters
Ammeters

Name: James Quinn   Date: 9-11-14

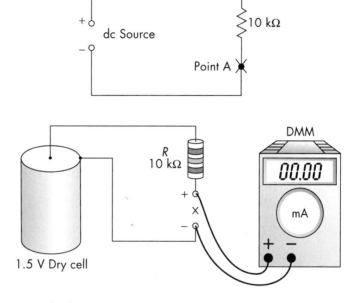

FIGURE 2-1 Standard schematic diagram (top) and pictorial diagram (bottom)

PROJECT PURPOSE To learn how to properly and safely measure dc circuit current by connecting simple circuits from schematic or pictorial diagrams and making current measurements.

PARTS NEEDED
- ☐ DMM
- ☐ 1.5-V dry cell
- ☐ Component Interconnection System (CIS) (e.g., Proto-Board, etc.)
- ☐ Resistor 10 kΩ

SAFETY HINTS Use proper function, range, and polarity to protect meter.

In this project, you should be aware of the following precautions and procedures related to connecting meters and making current measurements.

Safety Hints

1. Turn POWER OFF circuit into which ammeter will be connected.
2. BREAK THE CIRCUIT in the appropriate place and INSERT THE METER IN SERIES, observing the following rules.
3. Use the proper METER MODE/FUNCTION (dc current mode).

PART 1: Use and Care of Meters

4. On nonautoranging meters be sure the RANGE is high enough for the current to be measured. (Start with the HIGHEST range and work down, if appropriate.)
5. Be sure to OBSERVE PROPER POLARITY when measuring dc.
6. After meter is connected properly, TURN POWER ON AND TAKE READING.
7. Observe appropriate PERSONAL SAFETY rules.

PROCEDURE

1. Connect the circuit shown in Figure 2-1.

2. Set up the DMM to measure currents in the range of 0–1 mA dc.

    OBSERVATION Mode/Function switch = __A__. Range switch = _____.

3. Follow the safety rules outlined above. Break the circuit at point A and insert the meter.

    OBSERVATION Check: (_X_ Yes ___ No) Proper polarity observed? _____.

    Connected in series? __yes__. Proper range? _____.

    CONCLUSION Current will enter the meter through the (black, red) __red__ lead and will exit the meter through the (black, red) __Black__ lead.

4. Connect the 1.5-V cell to the circuit as shown and read the ammeter.

    OBSERVATION Measured current is: __.15__ mA.

    CONCLUSION What change(s) would need to be made in the meter setup if V were to be greatly increased or if R were to be greatly decreased? __We would have to change the range.__

PROJECT 2: Ammeters

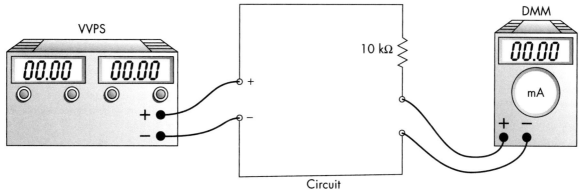

FIGURE 2-2 Pictorial diagram

PROJECT PURPOSE To provide practice in measuring dc circuit current and adjusting source voltage to obtain desired circuit current.

PARTS NEEDED
- [ ] DMM
- [ ] VVPS (dc)
- [ ] CIS
- [ ] Resistor 10 kΩ (1/2 W, or greater)

SAFETY HINTS Be sure power is off when making circuit changes and/or when connecting meters.

5. Disconnect the 1.5-V cell from the previous circuit in Figure 2-1. Set up the DMM to read current in the range of 0 to 10 mA. Have your instructor check your setup.

   OBSERVATION Setup OK. ▶ Instructor initial: _____

6. With power OFF, connect a variable-voltage power supply (VVPS) to the circuit so the direction of current through the circuit will be the same as it was when the 1.5-V cell was connected, Figure 2-2.

   OBSERVATION Setup OK. ▶ Instructor initial: _____

7. Set the voltage control on the VVPS to the zero output setting. Then, turn power supply on and SLOWLY adjust output until you measure 5 mA through the circuit.

   OBSERVATION Procedure OK. ▶ Instructor initial: _____

8. What is the voltage applied to the circuit when the VVPS is adjusted so that there is current flow of 2.5 mA?

   CAUTION: Remember the safety rules.

   OBSERVATION Voltage measures __24.9__ V.   Current measures __2.7__ mA.

Use and Care of Meters
Ohmmeters

PROJECT 3

Name: James Quinn   Date: 9-11-14

PROJECT PURPOSE To provide practice in using the ohmmeter function of a DMM.

In this project, you should be aware of the following precautions and procedures about using the ohmmeter to make resistance measurements.

⛔ Safety Hints

1. Turn POWER OFF and/or DISCONNECT circuit from power source.
2. ISOLATE COMPONENT being measured from the rest of the circuit, whenever possible, to prevent "sneak" paths.
3. Use the proper METER MODE/FUNCTION (dc and ohms).
4. On nonautoranging meters, be sure the RANGE is appropriate for the range of resistance anticipated.
5. Connect test probes across component or circuit to be tested and make the measurement.
6. BE SURE TO TURN SELECTOR SWITCH OFF of the "ohms" range to a high-voltage range position (e.g., 1,000 V) or to the "OFF" position, if your meter has it, when finished using the ohmmeter.

PROCEDURE

1. Obtain a 1,000-Ω and a 10,000-Ω resistor. Use proper procedures; measure and record the resistance value of each resistor using the ohmmeter.

   ⚠ OBSERVATION   The 1,000-Ω resistor measures 9,700 Ω.   The 10,000-Ω resistor measures 79,600 Ω.

   ⚠ CONCLUSION   Did the two resistors measure exactly 1,000 Ω and 10,000 Ω, respectively? No.
   What could have caused any differences? The exact resistor strength can vary slightly, the meter might also be off.

2. Obtain a kit of five unknown resistor values from your instructor. Using the ohmmeter, measure and record their values as numbered.

   OBSERVATION   Measured values are:

   R #1 = __7.4__ Ω.   R #4 = __.466__ kΩ.
   R #2 = __21.9__ Ω.   R #5 = __.577__ kΩ.
   R #3 = __1.48__ kΩ.

   ➤ Instructor initial: _____

   CONCLUSION   When the value of R to be measured is unknown, it is generally best to start by trying to measure on the (highest, lowest) __Highest__ R range and by then working (up, down) __down__ through the other ranges until the R value can be easily read.

3. If available, obtain several circuits from your instructor, some of which have continuity, some of which do not. Use the ohmmeter to identify the circuits that have continuity and those that do not have continuity.

   OBSERVATION   Continuity check: If there is continuity, fill in the blank with "yes," if not, fill in the blank with "no."

   Circuit #1 __yes__.   Circuit #4 __yes__.
   Circuit #2 __yes__.   Circuit #5 __no__.
   Circuit #3 __no__.

   CONCLUSION   When there was continuity, the ohmmeter reading indicated a (high, low) __high__ resistance reading.

   When there was NO continuity, the ohmmeter indicated __0__ ohms of resistance.

Summary
Use and Care of Meters

Name: James Quinn   Date: 9-11-14

Complete the following review questions, indicating the appropriate response by placing a check in the box next to the correct answer.

1. When preparing to measure an unknown dc voltage, the mode/function and range switches should be set at
   - ☐ dc mode, lowest V range
   - ☐ ac mode, lowest V range
   - ☒ dc mode, highest V range
   - ☐ ac mode, highest V range

2. The red test lead on a multimeter should be connected to the
   - ☒ positive input jack on the meter
   - ☐ negative input jack on the meter
   - ☐ neither of these

3. When measuring dc voltage, the meter's black test lead is normally connected to the more _____ point of the component or circuit being measured.
   - ☒ negative
   - ☐ positive

4. Precautions for preparing to measure current include
   - ☐ Circuit off; correct polarity; correct range; connect meter in parallel
   - ☒ Circuit off; correct polarity; correct range; connect meter in series
   - ☐ Circuit on; correct polarity; correct range; connect meter in parallel
   - ☐ Circuit on; correct polarity; correct range; connect meter in series

5. Precautions when using an ohmmeter to measure the resistance value of a resistor include
   - ☒ Power off; correct range; correct polarity; turn off ohms mode when through measuring R
   - ☐ Power off; correct range; turn off ohms mode when through measuring R
   - ☒ Power on; correct range; correct polarity; turn off ohms mode when through measuring R

6. Turning the meter off the ohms mode when through may prevent
   - ☒ Possibility of battery drainage; possibility of meter damage
   - ☐ Using wrong range; meter damaging a circuit

OHM'S LAW

PART 2

Objectives

You will connect several simple dc resistive circuits and make measurements and observations regarding how Ohm's law is applied to practical electrical circuits.

In completing these projects, you will use the color code, connect circuits, make measurements, perform calculations, draw conclusions, and be able to answer questions about the following items relating to the resistor color code and Ohm's law:

- Use of the resistor color code
- The relationship of current to voltage in simple resistive dc circuits
- The relationship of current to resistance in simple resistive dc circuits
- The relationship of power to voltage in simple resistive dc circuits
- The relationship of power to current in simple resistive dc circuits

Project/Topic Correlation Information

PROJECT	TEXT CHAPTER	SECTION	RELATED TEXT TOPIC(S)
4 Resistor Color Code Review and Practice	2	2-5	Resistor Color Code
5 Relationship of I and V with R Constant	3	3-1	The Relationship of Current to Voltage with Resistance Constant
6 Relationship of I and R with V Constant	3	3-1	The Relationship of Current to Resistance with Voltage Constant
7 Relationship of Power to V with R Constant	3	3-8	The Basic Power Formula
8 Relationship of Power to I with R Constant	3	3-8	The Basic Power Formula

Ohm's Law
Resistor Color Code Review and Practice

PROJECT 4

Name: James Quinn Date: 9-18-14

PROJECT PURPOSE To provide review and hands-on practice in using the resistor color code.

CHART 4-1

1st Color	2nd Color	3rd Color	4th Color	Ohm's Value	Tolerance Percent
Red	Violet	Yellow	Gold	27,000	5%
Brown	Black	Green	None	10,000,000	20%
Orange	White	Black	Gold	39	5%
Yellow	Violet	Orange	Silver	47,000	10%
Gray	Red	Brown	Gold	820	5%
Green	Brown	Black	Gold	51	5%
Blue	Red	Brown	Gold	620	5%
Green	Blue	Orange	Silver	56,000	10%

CHART 4-2

Ohm's Value	Colors		
100 Ω	Brown	Black	Black
12 Ω	Brown	Red	Black
1.0 Ω	Black	Brown	Red
13 kΩ	Brown	Orange	Black
2 MΩ	Red	Black	Green
91 Ω	White	Brown	Black

NOTE ▶ These charts are for practice only and therefore may have some values called for that are not standard available resistor values.

PART 2: Ohm's Law

PROCEDURE

1. List the ten colors used in the resistor color code to represent 0, 1, 2, 3, 4, 5, 6, 7, 8, and 9.

   ⚠ **OBSERVATION** Colors used in the resistor color code are as follows:

   0 = Black .   5 = Green .
   1 = Brown .   6 = Blue .
   2 = Red .   7 = Violet .
   3 = Orange .   8 = Grey .
   4 = Yellow .   9 = White .

2. List the other colors in the color code generally used to indicate resistor tolerance or used as special multipliers.

   ⚠ **OBSERVATION** Special colors are:
   Silver and Gold .

   ⚠ **CONCLUSION**
   The color used to indicate 5% tolerance is Gold .
   The color used to indicate 10% tolerance is Silver .
   The color used to indicate a 0.1 multiplier is Gold .
   The color used to indicate a 0.01 multiplier is Silver .

3. Fill in the resistance and tolerance values on Chart 4-1, as appropriate.

4. Fill in the colors that can be used to indicate the values in Chart 4-2, as appropriate.

5. Obtain a set of ten resistors having assorted values and tolerances and use the color code to determine their values and tolerances, as appropriate.

   ⚠ **OBSERVATION** Values and tolerances of resistors in the kit are as follows:

   R #1 = 120 Ω 5 % tolerance.   R #6 = 22 Ω 10 % tolerance.
   R #2 = 600 Ω 5 % tolerance.   R #7 = 2.7K Ω 5 % tolerance.
   R #3 = 330 Ω 10 % tolerance.   R #8 = 390 Ω 20 % tolerance.
   R #4 = 18 Ω 5 % tolerance.   R #9 = 47 Ω 10 % tolerance.
   R #5 = 1K Ω 20 % tolerance.   R #10 = 56 Ω 20 % tolerance.

   ➤ Instructor initial: _____

6. For precision resistors, a five-band color coding system is frequently used. List the meaning of each band in the observation column.

   ⚠ **OBSERVATION**
   Band #1 = Hundreds .   Band #4 = Multiplier .
   Band #2 = Tens .   Band #5 = Tolerance .
   Band #3 = Ones .

7. For the fifth band on these precision resistors, list the meaning of each color listed in the "Observation" section, as appropriate.

   ⚠ OBSERVATION Brown = _____%.   Blue = _____%.

   Red = _____%.   Violet = _____%.

   Green = _____%.

8. Fill in the resistance and tolerance values for the five-band precision resistors listed in Optional Chart 4-3, as appropriate.

Optional Chart

CHART 4-3 Special Precision Resistors

1st Color	2nd Color	3rd Color	4th Color	5th Color	Resistance Value (First 4 bands)	Tolerance Percent (5th band)
Brown	Brown	Black	Red	Brown		
Orange	Blue	Black	Red	Red		
Brown	Brown	Black	Orange	Brown		
White	Brown	Black	Orange	Red		
Red	Yellow	Black	Gold	Green		
Brown	Brown	Black	Silver	Blue		

Ohm's Law
Relationship of *I* and *V* with *R* Constant

Name: James Quinn   Date: 9-18-14

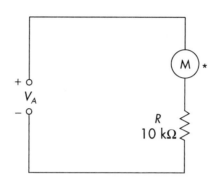

* Use best range to measure currents between 0–1 mA

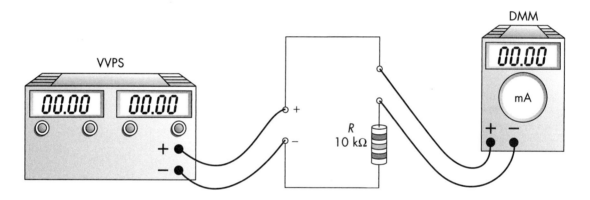

FIGURE 5-1 Standard schematic diagram (top) and pictorial diagram (bottom)

PROJECT PURPOSE To demonstrate Ohm's law and the direct relationship of current to voltage for a given resistance. To provide further practice in connecting circuits and making electrical measurements.

PARTS NEEDED
- ☐ DMM
- ☐ VVPS (dc)
- ☐ CIS
- ☐ 10-kΩ resistor

SAFETY HINTS Be sure power is off when connecting meters.

As you perform this project, remember that Ohm's law states that $I = \dfrac{V}{R}$.

PROCEDURE

1. Connect the initial circuit as shown in Figure 5-1.

2. Adjust V_A to obtain 1/2 scale deflection on the 1-mA range.

   OBSERVATION   Current is ___.986___ mA.

3. Measure the V_A.

   OBSERVATION V_A measures ___9.86___ V.

4. Use Ohm's law ($I = V/R$) and calculate I from the measured value of V and the indicated R value.

   OBSERVATION V_A = ___9.86___ V.   R = ___10,000___ Ω.

   CONCLUSION I calculated = ___.986___ mA.

5. Increase V_A to twice its original value and note the new current reading.

   OBSERVATION V_A = ___19.22___ V. I now = ___1.972___ mA.

   CONCLUSION   Doubling V_A caused I to: ___Double___. From this we conclude that with R constant (unchanged), current is directly proportional to ___voltage___.

6. Reduce V_A to 2 volts and note the new current reading.

   OBSERVATION I now = ___.2___ mA.

   CONCLUSION   Reducing V_A to 2 volts caused the current to (*increase, decrease*) ___decrease___ proportionately. This shows again that current stays "in step" (is directly proportional) with voltage when R is unchanged. This means that if R is held constant and V is increased, I will (*increase, decrease*) ___Increase___; if V is decreased, I will (*increase, decrease*) ___Decrease___.

Ohm's Law
Relationship of *I* and *R* with *V* Constant

Name: James Quinn   Date: 9-18-14

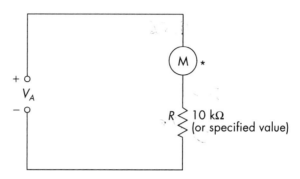

* Use best range to measure currents between 0–1 mA

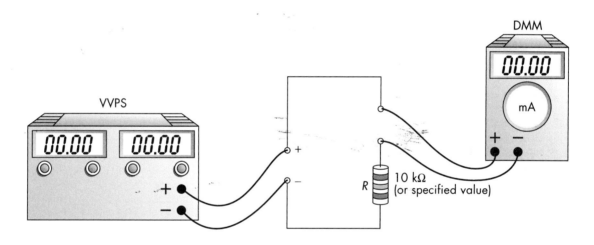

FIGURE 6-1 Standard schematic diagram (top) and pictorial diagram (bottom)

PROJECT PURPOSE To demonstrate Ohm's law and the inverse relationship of current to resistance for a given voltage. To provide continued practice in circuit connection and measurement.

PARTS NEEDED
- ☐ DMM
- ☐ VVPS (dc)
- ☐ CIS
- ☐ Resistors
  - 10 kΩ   27 kΩ
  - 47 kΩ   100 kΩ

SAFETY HINTS Be sure power is off when connecting meters or changing components.

PART 2: Ohm's Law

PROCEDURE

1. Connect the initial circuit as shown in Figure 6-1.

2. Adjust V_A to obtain 1 mA of current.

   OBSERVATION   Current is ___1___ mA.

3. Measure V_A and **be sure *not* to change** V_A for the rest of the steps in this section.

   OBSERVATION V_A measures ___9.83___ V.

4. Use Ohm's law ($R = V/I$) and calculate R from the measured values of V and I.

   OBSERVATION V = ___9.83___ V. I = ___1___ mA.

   CONCLUSION R calculated = ___10k___ Ω.

5. Remove the 10-kΩ resistor from the circuit, replace it with a 100-kΩ resistor, and note the new current reading.

   OBSERVATION R now = ___100,000___ Ω. I now = ___.1___ mA.

   CONCLUSION   Keeping V_A constant at ___9.83___ volts and increasing R by ten times to a value of ___100,000___ ohms caused the current to (*increase, decrease*) ___decrease___ to one ___tenth___ of its original value. From this we conclude that I is inversely proportional to R. This means that if R is increased, I will (*increase, decrease*) ___decreases___ proportionately; if R is decreased, I will (*increase, decrease*) ___Increases___ by the same factor as R was decreased.

6. Remove the 100-kΩ resistor from the circuit, replace it with a 47-kΩ resistor, and record the new current reading.

   NOTE ▶ Keep V_A the same as it was.

   OBSERVATION V_A = ___9.83___ V. I = ___.21___ mA.
   R = ___47___ kΩ.

   CONCLUSION   The circuit resistance for this step is *approximately* (*1/4, 1/2*) ___½___ that of the previous step; the applied voltage is the same; and the resulting current is approximately (*2, 4*) ___2___ times that of step 5. This again tends to prove that I is (*directly, inversely*) ___Inversly___ proportional to R, with V held constant.

Optional Step

7. Remove the 47-kΩ resistor from the circuit and replace it with a 27-kΩ resistor. Record the new current reading.

   OBSERVATION V_A = __9.83__ V. I = __.36__ mA.
   R = __27__ kΩ.

   CONCLUSION   Did the circuit current change from the previous step? __yes__. Was the R larger or smaller than in the previous step? __Smaller__. In your own words, explain what this data proves: __I is inversely proportional to R.__

Ohm's Law
Relationship of Power to *V* with *R* Constant

Name: James Quinn   Date: 9-18-14

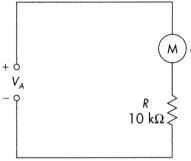

* Use best range to measure currents between 0–1 mA

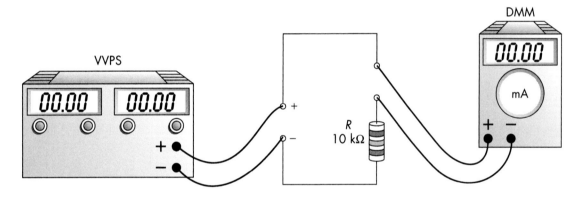

FIGURE 7-1 Standard schematic diagram (top) and pictorial diagram (bottom)

PROJECT PURPOSE To demonstrate the use of the power formula, which shows that power is related to voltage squared for a given resistance. To provide continued practice in circuit connection and measurement.

PARTS NEEDED
- ☐ DMM
- ☐ VVPS (dc)
- ☐ CIS
- ☐ 10-kΩ resistor

SAFETY HINTS Be sure power is off when connecting meters.

PROCEDURE

1. Connect the initial circuit as shown in Figure 7-1.
2. Adjust V_A to obtain 0.~~5~~ 7 mA of current and measure V_A.

   ⚠ OBSERVATION   I = ~~.~~.7 mA. V_A = 15 m V.

3. Use the $V \times I$ power formula and calculate the power dissipated by R.

   ⚠ OBSERVATION   V = .015 ~~m~~ V.   I = .7 mA.

   ⚠ CONCLUSION   P calculated = 10.5 mW.

4. Change V_A to obtain 1 mA of current. Measure V_A and calculate P using the $V \times I$ formula.

   ⚠ OBSERVATION   V now = 18 m V.   I now = 1.2 mA.

   ⚠ CONCLUSION   P calculated now = 21 mW. Doubling V caused the current to increase by how many times? 2 . Thus, the product of $V \times I$ increased 2 times (when we doubled the voltage and kept R constant). From this we conclude that power is proportional to the _____ of the voltage when R is not changed.

5. Calculate the power for the measured values of step 4 using the $P = V^2/R$ formula.

   ⚠ OBSERVATION   V = ~~~~ 18 V.   R = 10 k Ω.

   ⚠ CONCLUSION   P calculated = _____ mW.

6. Change V_A to 2.5 volts. Measure the appropriate voltage and current values and calculate P by both of the above formulas.

   ⚠ OBSERVATION   V now = 2.55 V.   I now = 203.8 mA.

   ⚠ CONCLUSION   P calculated by both methods is approximately 519.7 mW. Compared to the 10-volt V_A condition, we now have (1/2, 1/4, 1/8) 1/8 the V_A and (1/4, 1/8, 1/16, 1/32) _____ the power dissipation. This again illustrates that P is related to the (square, √) √ of the voltage when R remains unchanged.

Optional Steps

7. Assume a V_A value of 4 volts. Calculate and predict what the circuit current and power would be for that value of applied voltage.

   ⚠ OBSERVATION   Predicted I = _____ mA.   Predicted P = _____ mW.

8. Change the circuit V_A to 4 volts, measure I, and calculate P.

   ⚠ OBSERVATION   Measured I = _____ mA.   Calculated P = _____ mW.

▲ **CONCLUSION** Was the current with 4 volts applied voltage higher or lower than when there were 2.5 volts applied? _____. Was the calculated power higher or lower? _____. How many times higher or lower? _____. Is the change coherent with the concept of power being proportional to V^2? _____.

Ohm's Law
Relationship of Power to *I* with *R* Constant

PROJECT 8

Name: James Quinn   Date: 9-18-18

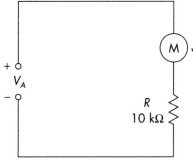

* Use best range to measure currents between 0–1 mA

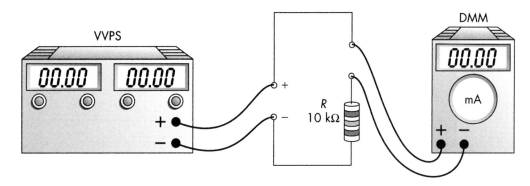

FIGURE 8-1 Standard schematic diagram (top) and pictorial diagram (bottom)

PROJECT PURPOSE To verify the fact that power is related to the current squared for a given resistance. To provide continued practice in circuit connection and measurement.

PARTS NEEDED
- ☐ DMM
- ☐ VVPS (dc)
- ☐ CIS
- ☐ 10-kΩ resistor

SAFETY HINTS Be sure power is off when connecting meters.

PROCEDURE

1. Connect the initial circuit as shown in Figure 8-1.

2. Adjust V_A to 7 volts and note the current.

    ⚠ OBSERVATION $I =$ __.7__ mA.

3. Use the measured values of V and I and calculate P by the formula $P = V \times I$.

    ⚠ OBSERVATION $V =$ __7__ V. $I =$ __.7__ mA.

    ⚠ CONCLUSION P calculated = __4.9__ mW.

4. Decrease V_A until I is 1/2 its original value and calculate P using both the $V \times I$ and $I^2 R$ formulas.

    ⚠ OBSERVATION V now = __3.5__ V. I now = __.35__ mA.

    ⚠ CONCLUSION P calculated by both methods is approximately __1.225__ mW. This is (1/8, 1/4, 1/2) __1/4__ the power dissipated when I was double the value of the current for this step. From this we can conclude that P is proportional to I^2 when R in the circuit is unchanged. This means that if the circuit current has doubled, the power dissipation has increased (1, 2, 3, 4) __4__ times; or, if circuit current were decreased to one-third its original value, the power must decrease to (1/3, 1/6, 1/9) __1/9__ its original value.

Optional Step

5. Change V_A until it is approximately one-third the original 7-volt value (approximately 2.33 volts). Make appropriate measurements and calculations to fill in the blanks in the "Observation" section.

    ⚠ OBSERVATION V_A now = __2.33__ V. $V \times I$ now = __540__ mW.

    I now = __.23__ mW V^2/R now = __.5__ mW.

    ⚠ CONCLUSION   Is the power approximately one-ninth that when 7 volts was applied? __yes__. Does this verify that power is related to the square of the circuit current with R constant? __yes__.

Story Behind the Numbers
Ohm's Law

Name: _____ Date: _____

Procedure

NOTE When performing the procedure steps, you have the options of using a calculator or using an Excel spreadsheet program "worksheet" for any required calculations. You may also use Excel for creating tables and for generating graphs.

1. Connect the circuit as shown.

2. Temporarily turn off the power supply and insert a DMM in the current mode to measure circuit current, using the appropriate polarity and safety rules while making the connections. Measure the voltage across the circuit resistor, as appropriate.

3. Turn on the power supply and adjust for 10 volts across R (which is the same as V_A in this case). Observe the circuit current reading. Record the *Given* parameter values and the *Measured* circuit current in the appropriate locations on the data table provided in this project, or set up an Excel program worksheet similar to the table.

4. Make appropriate changes in the values of R and the circuit applied voltage to complete the measured circuit current value for each set of conditions shown in the table.

5. After performing all the circuit changes, measurements, and logging of data needed to fill in the table, use the data to create the two line graphs described in A and B.

    NOTE ▶ For basic information on creating graphs (charts) with Excel, see Appendix B.

    A. Line graph 1 should show the relationship of current to resistance value when applied voltage is 10 volts. Label the **x-axis** to show the three different resistor values used. Show the first $R = 10$ kΩ, the second $R = 27$ kΩ, and the third $R = 47$ kΩ. The **y-axis** should be labeled "Current Value."

    B. Line graph 2 should also show the relationship of current to resistance; however, this time use the data acquired when applied voltage was set at 20 volts. Once again, label the **x-axis** to show the three resistor values used. The **y-axis** will still show the current values, and this axis should again be labeled "Current Value."

6. After completing the data table and producing the required graphs, answer the Analysis Questions and create the brief Technical Lab Report in order to complete the project.

Data Table

Given Values		Measured Current
V Applied Value	**Value of R Used**	**Value of Circuit Current (in mA)**
10 V	10 kΩ	1 mA
10 V	27 kΩ	.37 mA
10 V	47 kΩ	.21 mA
20 V	10 kΩ	2 mA
20 V	27 kΩ	.74 mA
20 V	47 kΩ	.42 mA

Analysis Questions

NOTE Answers to these Analysis Questions should be clearly numbered and documented on separate sheets of paper with your name and the date at the top of each page. These answer sheets are to be turned in with the rest of the project documentation, as appropriate.

1. Look at the data table and the two graphs you have created. Give your analysis of the relationship of circuit current to resistance value for any given applied voltage.

2. Study the two graphs you made that used the same R values but a different applied voltage for each graph. What characteristic or characteristics on these two graphs show a definite ratio relationship? What parameter(s) show this ratio relationship, and what is the ratio that is apparent?

3. As an added bit of analysis, calculate the total circuit power provided by the source for each circuit R at each of the applied voltages used in this project.

   HINT ▶ If you are using Excel, you might just add a couple of columns to your original chart (V_A value *number only* in one column and the *appropriate Excel formula* in an adjacent column's cells where you want the power value results to appear).

4. In a logical manner, describe any special power value ratios that you can observe in the preceding step.

5. Summarize the specific Ohm's law inverse and direct relationships you have observed in performing this project.

Technical Lab Report

Write a brief technical lab report summarizing the technical facts learned from this project. The report should be organized to provide the following:

1. An introductory paragraph describing the type of circuit being analyzed and the key parameters that will be discussed relating to this circuit.

2. A section describing the most important characteristics of this type of circuit that were shown via the collected data in the tables and graphs.

3. Any special facts or characteristics about this type of circuit that were highlighted in answering the Analysis Questions.

4. A practical example of how the information learned in this project might help you in operating, troubleshooting, error analysis, or adjusting a circuit of this type in your home setting, in your training program setting, or in a job setting in the real world.

5. A summary statement listing the most positive aspects of the project and any parts of the project that were difficult because of equipment problems or unclear instructions. Include areas that might be improved.

SUMMARY
Ohm's Law

Name: _____ Date: _____

Complete the following review questions, indicating the appropriate response by placing a check in the box next to the correct answer.

1. If V increases and R remains the same, then I will
   - ☐ increase
   - ☐ decrease
   - ☐ remain the same

2. If I increases and R remains the same, then V must have
   - ☐ increased
   - ☐ decreased
   - ☐ remained the same

3. If R increases and V remains the same, then I will
   - ☐ increase
   - ☐ decrease
   - ☐ remain the same

4. If V is doubled and R is halved, then I will
   - ☐ double
   - ☐ halve
   - ☐ quadruple
   - ☐ remain the same

5. If V is doubled and R remains the same, then P will
   - ☐ double
   - ☐ halve
   - ☐ quadruple
   - ☐ remain the same

6. If I is halved and R remains the same, then P will
   - ☐ double
   - ☐ halve
   - ☐ quadruple
   - ☐ decrease to one-quarter
   - ☐ remain the same

7. If V is doubled and R is halved, then P will
   - ☐ decrease 4 times
   - ☐ increase 8 times
   - ☐ increase 16 times
   - ☐ remain the same

8. Increasing the voltage applied to a circuit will cause
   a. Current to
      - ☐ increase
      - ☐ decrease
      - ☐ remain the same
   b. Resistance to
      - ☐ increase
      - ☐ decrease
      - ☐ remain the same
   c. Power dissipated to
      - ☐ increase
      - ☐ decrease
      - ☐ remain the same

9. Decreasing the resistance in a circuit will cause
   a. Current to
      □ increase   □ remain the same
      □ decrease
   b. Voltage applied to
      □ increase   □ remain the same
      □ decrease
   c. Power dissipated to
      □ increase   □ remain the same
      □ decrease

10. In an electrical circuit
    a. Current is directly proportional to
       □ V □ R
    b. And inversely proportional to
       □ V □ R
    c. While power is proportional to the square of the
       □ V □ R

SERIES CIRCUITS

PART 3

Objectives

You will connect several dc resistive series circuits and make measurements and observations regarding their important electrical characteristics.

In completing these projects, you will connect circuits, make measurements, perform calculations, draw conclusions, and be able to answer questions about the following items related to series circuits:

- Total resistance
- Voltage distribution
- Circuit current
- Power dissipation(s)
- Effects of opens
- Effects of shorts
- Application of Ohm's law
- Application of Kirchhoff's voltage law

Project/Topic Correlation Information

PROJECT		TEXT CHAPTER	SECTION	RELATED TEXT TOPIC(S)
9	Total Resistance in Series Circuits	4	4-2	Resistance in Series Circuits
10	Current in Series Circuits	4	4-1	Definition and Characteristics of a Series Circuit
11	Voltage Distribution in Series Circuits	4	4-3	Voltage in Series Circuits
12	Power Distribution in Series Circuits	4	4-5	Power in Series Circuits
13	Effects of an Open in Series Circuits	4	4-6	Effects of Opens in Series Circuits and Troubleshooting Hints
14	Effects of a Short in Series Circuits	4	4-7	Effects of Shorts in Series Circuits and Troubleshooting Hints

Series Circuits
Total Resistance in Series Circuits

PROJECT 9

Name: James Quinn   Date: 10-2-14

NOTE ▶ For this project, DO NOT connect power to this circuit!

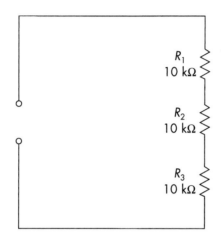

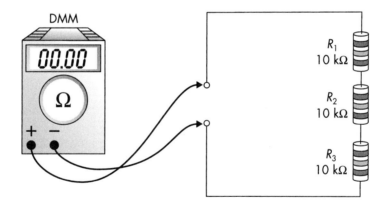

FIGURE 9-1 Standard schematic diagram (top) and pictorial diagram (bottom)

PROJECT PURPOSE To confirm the series circuit total resistance (R_T) formula by varying circuit conditions and making resistance measurements.

PARTS NEEDED
- ☐ DMM
- ☐ CIS
- ☐ Resistors
  - 10 kΩ (3)   47 kΩ
  - 27 kΩ   100 kΩ (2)

 SAFETY HINTS DO NOT use a power supply for this project! (No power is to be applied when using an ohmmeter.)

39

PROCEDURE

1. Connect the initial circuit shown in Figure 9-1.

   ! CAUTION: Do not connect power to the circuit for this project!

2. Use an ohmmeter and measure the total resistance of the circuit.

   OBSERVATION R total = __30 k__ Ω.

3. Use an ohmmeter, measure the resistance of each individual resistor, and record your observations.

   OBSERVATION R_1 = __10 k__ Ω. R_3 = __10 k__ Ω.

   R_2 = __10 k__ Ω.

4. Add the resistances of the individual resistors and note the result.

   OBSERVATION $R_1 + R_2 + R_3$ = __30 k__ Ω.

   CONCLUSION   The total resistance of a series circuit equals the (*product*, *sum*) __Sum__ of all the individual resistances; therefore, R_T = __$R_1 + R_2 + R_3$__.

5. Predict what the new R_T would be if R_1 were changed to a 27-kΩ resistor.

   OBSERVATION   Predicted R_T = __47__ Ω.

6. Change R_1 to 27 kΩ and measure the new R_T.

   OBSERVATION   New R_T = __47__ Ω.

   CONCLUSION   Changing any element's resistance in a series circuit while the rest of the elements are unchanged will cause the circuit's total resistance to (*remain the same*, *change*) __change__. If any element's R increases, then R_T will (*increase*, *decrease*) __Increase__. If any element's R decreases, then R_T will (*increase*, *decrease*) __decrease__.

Optional Step

7. Change the circuit so that R_1 = 47 kΩ, R_2 = 100 kΩ, and R_3 = 100 kΩ. Predict R_T's value; then measure R_T to verify your prediction.

   OBSERVATION   Predicted R_T = _____ kΩ.   Measured R_T = _____ kΩ.

   CONCLUSION   Did R_T equal the sum of the individual resistances in this case? _____. What might cause a difference in predicted and measured values? _____

Series Circuits
Current in Series Circuits

Name: James Quinn   Date: 10-2-14

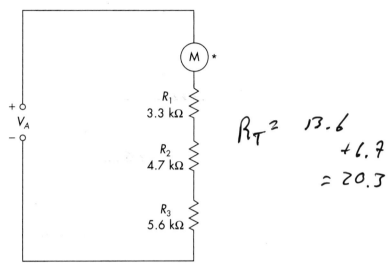

$R_T = 13.6$
$+ 6.7$
$= 20.3$

* Use best range to measure currents between 0–1 mA

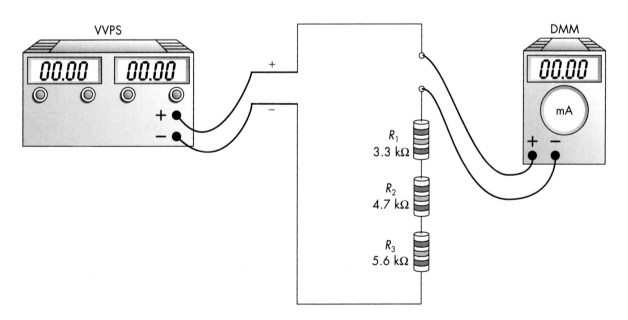

FIGURE 10-1 Standard schematic diagram (top) and pictorial diagram (bottom)

PROJECT PURPOSE To verify that the current throughout a series circuit is the same current via hands-on experience in measuring current at various points throughout a series circuit.

PARTS NEEDED
- DMM
- VVPS (dc)
- CIS
- Resistors: 3.3 kΩ, 4.7 kΩ, 5.6 kΩ, 10 kΩ

SAFETY HINTS Be sure power is off when connecting meters or changing components.

PROCEDURE

1. Connect the initial circuit shown in Figure 10-1.

2. Apply 9.5 volts to the circuit and note the current.

   OBSERVATION Number of paths for current in circuit = ___One___.
   Current = ___.6___ mA.

3. Swap positions of R_1 and the current meter and note the current reading.

   OBSERVATION Current reading is now ___.6___ mA.

4. Swap positions of the current meter and each of the remaining resistors and note the current reading each time.

   OBSERVATION Current reading in every case was ___.6___ mA.

   CONCLUSION No matter where the current meter was placed in the circuit, the current reading was the same. This indicates that the current through all parts of a series circuit is the ___Same___ current.

5. Change R_1 to a 10-kΩ resistor. Move the meter to various spots in the circuit and note the current reading.

   OBSERVATION Current reading in all cases was ___.4___ mA.

   CONCLUSION Increasing any R in a series circuit affects the current through all parts of the circuit. Would changing R by decreasing it cause I to change? ___yes___. Increase or decrease? ___Increase___.

6. Now change V_A to 19 volts. Move the meter to various spots in the circuit and note the current reading.

   OBSERVATION Current reading in all cases was ___.9___ mA.

   CONCLUSION Changing V_A for the series circuit caused I to change through all parts of the circuit. It changed by (*a different, the same*) ___the same___ amount in all parts because the current through all parts of a series circuit is (*a different, the same*) ___the same___ current. This is because there is only ___One___ path for current through series elements.

Series Circuits
Voltage Distribution in Series Circuits

Name: James Quinn   Date: 10-2-14

38 kΩ

FIGURE 11-1 Standard schematic diagram (top) and pictorial diagram (bottom)

PROJECT PURPOSE   To demonstrate the proportional relationship of Rs and Vs in series circuits through circuit measurements and calculations.

PARTS NEEDED
- ☐ DMM
- ☐ VVPS (dc)
- ☐ CIS
- ☐ Resistors
  - 1 kΩ
  - 10 kΩ
  - 27 kΩ

SAFETY HINTS   Be sure power is off when connecting meters.

44 PART 3: Series Circuits

(handwritten at top:) $R_T = 38\ K\Omega$, $I_T = .2\ mA$

PROCEDURE

1. Connect the initial circuit shown in Figure 11-1.

2. Apply 9.5 volts to the circuit and measure each of the individual voltage drops. Also calculate the circuit current.

   ⚠ OBSERVATION $V_A =$ __9.5__ V. $V_{R_3} =$ __6.75__ V.

   $V_{R_1} =$ __.25__ V. I calculated = __.2__ mA.

   $V_{R_2} =$ __2.5__ V.

   ⚠ CONCLUSION   Since the I is the same through all the resistors, the voltage drop across any given resistor is directly related to its R compared to the total circuit (*number of Rs, resistance*) __resistance__.

3. Add all the individual voltage drops and note the sum.

   ⚠ OBSERVATION $V_{R_1} + V_{R_2} + V_{R_3}$ equals __9.5__ V or V _____.

   ⚠ CONCLUSION   In essence, Kirchhoff's voltage law states that the arithmetic sum of voltage drops around any circuit closed loop must equal V applied. Does it? __yes__.

4. Calculate what fraction of the applied voltage is dropped by each of the resistors. Express your answer as a fraction. (Example: 1/38, 10/38, etc.)

   ⚠ OBSERVATION   *Example:*

   $V_{R_1} =$ __1/38th__ V_A. $V_{R_3} =$ __27/38__ V_A.

   $V_{R_2} =$ __10/38__ V_A.

   ⚠ CONCLUSION   Each resistor dropped the same fraction of V applied as its __resistor__ value is of the total _____.

5. Compute the fractional relationship of V_{R_1} to V_{R_2} and V_{R_3}. Express answers as fractions.

   ⚠ OBSERVATION   *Example:*

   $V_{R_1} =$ __1/10th__ V_{R_2}. $V_{R_1} =$ __1/27th__ V_{R_3}.

   ⚠ CONCLUSION   Because I is the same through all elements in a series circuit and since $V =$ __I__ $\times$ __R__, the voltage drops across the resistors are related to each other by the same factor as their __current__.

6. Predict what value V_{R_2} would be if V_A were 19 volts. Change V_A to 19 volts and measure V_{R_2}.

   ⚠ OBSERVATION V_{R_2} predicted = __2__ V. V_{R_2} measured = __2__ V.

   ⚠ CONCLUSION V_{R_2} is now what fraction of V_A? __10/38__. Is this the same fraction as when $V_A = 9.5$ volts? __10/38__. Changing V (*does, does not*) __does not__ change the distribution percentages of V_A. If one of the Rs were changed, would the distribution percentages change? __yes__. We may conclude that in series circuits, the largest R will drop the (*least, most*) __Most__ voltage, and the smallest R the (*least, most*) __Least__ voltage.

7. Use the voltage divider rule and calculate V_{R_1} and V_{R_3} assuming $V_T = 19$ V.

   ⚠ **OBSERVATION** V_{R_1} calculated = __.5__ V. V_{R_3} calculated = __13.5__ V.

   $$(V_x = \frac{R_x}{R_T} \times V_T)$$

8. Measure V_{R_1} and V_{R_3} with 19 V applied to circuit.

   ⚠ **OBSERVATION** V_{R_1} measured = __.5__ V. V_{R_3} measured = __13.5__ V.

   ⚠ **CONCLUSION**   Do the measured values for V_{R_1} and V_{R_3} confirm the calculations using the voltage divider rule? (*Yes, No*) __yes__.

Optional Steps

9. Turn off the power supply and replace R_3 with a 100-kΩ resistor.

10. Adjust voltage applied to the circuit to approximately 29 volts and measure the resistor voltage drops.

    ⚠ **OBSERVATION** V_{R_1} = _____ V. V_{R_3} = _____ V.
    V_{R_2} = _____ V.

    ⚠ **CONCLUSION**   Is V_{R_3} equal to about ten times V_{R_2}? _____? Is V_{R_2} about ten times greater than V_{R_1}? _____. Is the ratio of V_{R_3}'s voltage to V_{R_1}'s voltage about equal to their R ratio? _____. What might cause the ratios discussed to not be exactly 10:1 in each case?

    _____
    _____

Series Circuits
Power Distribution in Series Circuits

PROJECT 12

Name: James Quinn   Date: 10-9-14

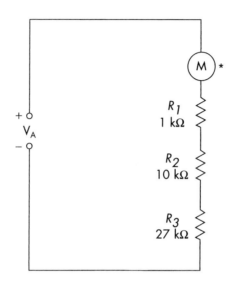

* Use best range to measure currents between 0–1 mA.

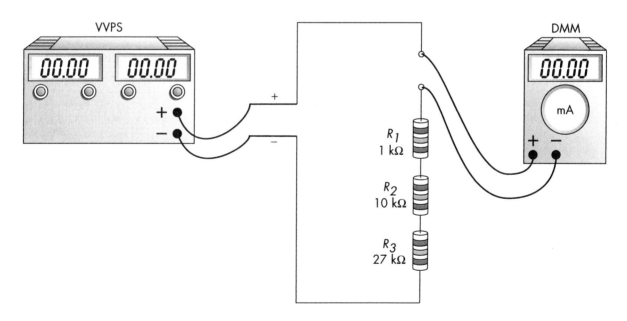

FIGURE 12-1 Standard schematic diagram (top) and pictorial diagram (bottom)

PROJECT PURPOSE To illustrate that power distribution in a series circuit is directly related to resistance distribution and that total power equals the sum of all the individual power dissipations.

48 PART 3: Series Circuits

PARTS NEEDED
- ☐ DMM
- ☐ VVPS (dc)
- ☐ CIS
- ☐ Resistors
  - 1 kΩ   27 kΩ
  - 10 kΩ

STOP SAFETY HINTS Be sure power is off when connecting meters.

$R_T = 38\ k\Omega$

PROCEDURE

1. Connect the initial circuit shown in Figure 12-1.

2. Apply 19 volts (V_A) to the circuit. Measure the current and the individual voltage drops and calculate the power dissipated by each of the resistors.

   OBSERVATION $I =$ __.5__ mA. $V_{R_2} =$ __5__ V.
   $V_{R_1} =$ __.5__ V. $V_{R_3} =$ __13.5__ V.

   CONCLUSION P_{R_1} calculated = __.25__ mW. P_{R_3} calculated = __6.75__ mW.
   P_{R_2} calculated = __2.5__ mW.

   We may conclude that since the I is the same through all resistors, the I^2R or (*voltage, power*) __Power__ dissipated by each resistor is directly related to its (*size, resistance*) __Resistance__ value. Furthermore, the power distribution throughout the circuit is the same as the (I, R) __I__ distribution. This means that the largest value R will dissipate the (*most, least*) __Most__ amount of power; the smallest R, the (*most, least*) __Least__ power.

3. Add all the individual power dissipations and note the sum. Also, calculate P_T by the formula:
   $P_T = V_T \times I_T$.

   OBSERVATION $P_{R_1} + P_{R_2} + P_{R_3}$ equals __9.5__ mW. $V_T \times I_T =$ __9.5__ mW.

   CONCLUSION The total power in a series circuit is equal to the (*product, sum*) __Sum__ of all the individual power dissipations.

4. What ratio does P_{R_1} have to P_{R_3}? Express as a ratio. (For example: 1:10, 2:5, etc.)

   OBSERVATION Ratio = __.25 : 6.75__.

   CONCLUSION The ratios of power dissipated by two resistors in a series circuit is the same as their __Voltage__ ratio.

5. Indicate what would happen to P_T and to the individual power dissipations if V_A were cut in half.

   OBSERVATION $P_T =$ __Half__ as much as before. Individual Ps would also be __Half__ original.

   CONCLUSION For a given R, decreasing V to one-half will also cause the circuit current to (*increase, decrease*) __decrease__ to __half__. Therefore, the product of $V \times I$ will be (1/2, 1/4) __1/4__ the original value. If Rs remain unchanged but V is changed, the percentage of P_T dissipated by any given R will (*change, not change*) ___change___ but the actual value of power dissipated will (*change, not change*) ___not change___.

Optional Step

6. Change the V_A to a value of 9.5 volts. Measure and calculate values, as required to fill in the blanks in the "Observation" section.

⚠ OBSERVATION

$V_A =$ _____ V. $V_{R_2} =$ _____ V.

$I =$ _____ mA. $P_{R_2} =$ _____ mW.

$P_T =$ _____ mW. $V_{R_3} =$ _____ V.

$V_{R_1} =$ _____ V. $P_{R_3} =$ _____ mW.

$P_{R_1} =$ _____ mW.

⚠ CONCLUSION When V_A was reduced to half its original value, current decreased to _____ its original value and total power dissipated by the circuit decreased to _____ its original value. Does the power dissipated by each resistor also change by this same factor? _____. Does this verify the data in step 5? _____.

Series Circuits
Effects of an Open in Series Circuits

PROJECT 13

Name: James Quinn   Date: 10-9-14

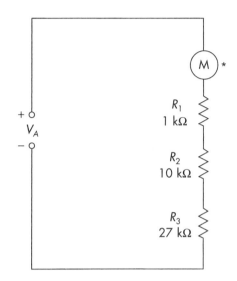

* Use best range to measure currents between 0–1 mA

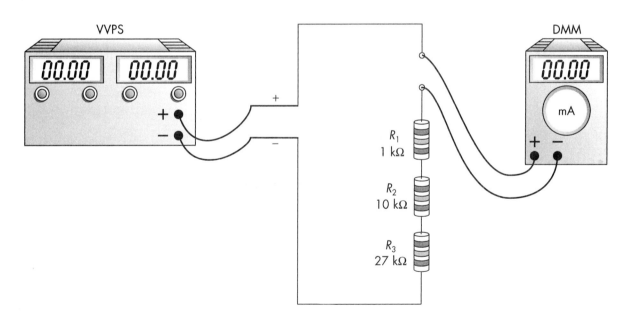

FIGURE 13-1 Standard schematic diagram (top) and pictorial diagram (bottom)

PROJECT PURPOSE To provide hands-on experience regarding circuit parameter changes that occur when an open develops in a series circuit and to verify that voltage applied appears across the open portion of the circuit.

PARTS NEEDED
- ☐ DMM
- ☐ VVPS (dc)
- ☐ CIS
- ☐ Resistors
  - 1 kΩ 27 kΩ
  - 10 kΩ

SAFETY HINTS Be sure power is off when connecting meters.

$R_T = 38\ k\Omega$

PROCEDURE

1. Connect the initial circuit as shown in Figure 13-1.

2. Adjust V applied to 19 volts. Measure and record the current and individual voltage drops.

   OBSERVATION
   $I = $ .5 mA. $V_{R_2} = $ 5 V.
   $V_{R_1} = $ .5 V. $V_{R_3} = $ 13.5 V.

3. To simulate an R becoming open in this series circuit, remove R_2 and leave the circuit open between R_1 and R_3. Measure and record the circuit current and the individual voltage drops.

   OBSERVATION
   $I = $ 0 mA. $V_{R_2} = $ 0 V. (across open)
   $V_{R_1} = $ 0 V. $V_{R_3} = $ 0 V.

   CONCLUSION Since in a series circuit there is only one path for current flow, opening any element within the series circuit will cause (continuity, _discontinuity_) _____. The R_T of the circuit then appears to be infinitely (_high_, low) _____. The voltage drop across R_1 was __0__ volts because, with zero current, the $I \times R$ drop must be __0__. We conclude that if any part of a series circuit opens, R_T will (_increase_, decrease) __increase__ to __infinite__; I_T will (increase, decrease) __decrease__ to __zero__. The voltage drops across the unopened elements will (increase, _decrease_) __decrease__ to __zero__, and the potential difference across the open portion of the circuit will (increase, decrease) __increase__ to __infinite__.

4. Assume R_2 was replaced in the circuit and R_3 removed. List the predicted results in the "Observation" section.

   OBSERVATION
   $I = $ 0 mA. $V_{R_2} = $ 0 V.
   $V_{R_1} = $ 0 V. $V_{R_3} = $ 0 V.

5. Make the change suggested in step 4, make appropriate measurements, and note the results.

⚠ OBSERVATION $I =$ __0__ mA. $V_{R_2} =$ __0__ V.

$V_{R_1} =$ __0__ V. $V_{R_3} =$ __0__ V.

⚠ CONCLUSION Do the results of these measurements verify the conclusions of step 3? __yes__.
The main difference noted in the parameters for this step compared to step 3 is that V applied now appears across R __3__, rather than R __2__.

Series Circuits
Effects of a Short in Series Circuits

PROJECT 14

Name: James Quinn   Date: 10-9-14

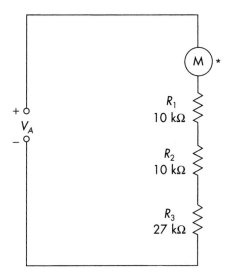

* Use best range to measure currents between 0–1 mA

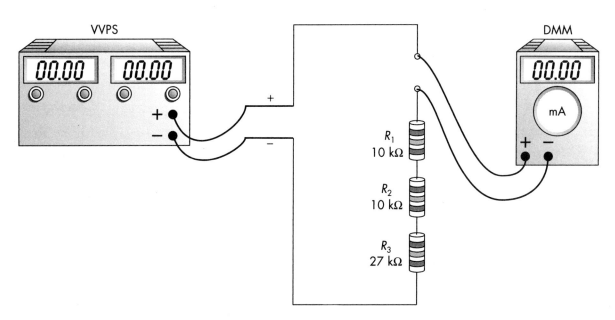

FIGURE 14-1 Standard schematic diagram (top) and pictorial diagram (bottom)

PROJECT PURPOSE To provide hands-on experience regarding circuit parameter changes that occur when a short develops in part of a series circuit.

PARTS NEEDED
- ☐ DMM
- ☐ VVPS (dc)
- ☐ CIS
- ☐ Resistors
  10 kΩ (2)   27 kΩ

STOP SAFETY HINTS DO NOT SHORT OUT ALL THREE RESISTORS AT ONCE! Be sure power is off when connecting meters.

PROCEDURE

1. Connect the initial circuit as shown in Figure 14-1.

2. Adjust V applied to 20 volts. Measure and record the current and individual voltage drops.

   ⚠ OBSERVATION $I =$ __.4__ mA. $V_{R_2} =$ __4.2__ V.
   $V_{R_1} =$ __4.2__ V. $V_{R_3} =$ __11.4__ V.

3. To simulate an R "shorting out," remove R_2 and replace it with a jumper wire. Measure and note the circuit I and individual voltage drops with the shorted element replacing R_2.

   ⚠ OBSERVATION $I =$ __.5__ mA. $V_{R_2} =$ __0__ V.
   (across short)
   $V_{R_1} =$ __5.4__ V. $V_{R_3} =$ __14.5__ V.

⚠ CONCLUSION   Shorting out R_2 has caused the circuit R_T to (increase, *decrease*) __decrease__ to __37__ ohms. This then caused I_T to (*increase*, decrease) __Increase__. The new (higher, lower) __Higher__ current caused the $I \times R$ drops across the unshorted elements (Rs) in the circuit to (increase, decrease) __Increase__. The R between the shorted terminals is effectively __0__ ohms. Therefore, the $I \times R$ drop across the shorted element or section of a series circuit will (increase, decrease) __decrease__ to __zero__. We conclude that if any part of a series circuit shorts, R_T will (increase, *decrease*) ~~decrease~~ ; I_T will (increase, decrease) __Increase__; the voltage drops across the unshorted elements will (*increase*, decrease) __Increase__; and the V across the shorted element or section of the circuit will (increase, *decrease*) __decrease__ to __zero__. The total power supplied to the circuit will (increase, decrease) __Increase__ with the shorted condition.

Story Behind the Numbers
Series Circuits

Name: James Quinn   Date: 10-9-14

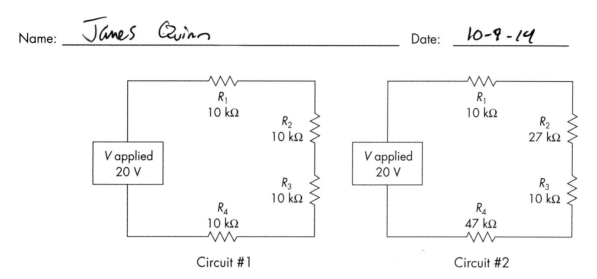

Circuit #1   Circuit #2

Procedure

NOTE When performing the procedure steps, you have the options of using a calculator or using an Excel spreadsheet program "worksheet" for any required calculations. You may also use Excel for creating tables and for generating graphs.

1. Connect Circuit 1. Make measurements of component values and circuit parameters for Circuit 1 as called for in Table 1 and fill in the Table 1 data, as appropriate.

   ! CAUTION: Be sure power source is disconnected from circuit when making R measurements!

2. Connect Circuit 2. Make measurements of component values and circuit parameters for Circuit 2 as called for in Table 1 and fill in the Table 1 data, as appropriate.

   ! CAUTION: Be sure power source is disconnected from circuit when making R measurements!

3. Create a line graph for Circuit 1 paramenters. For the Circuit 1 graph, label the **x-axis** of the graph with the identifiers: $R_1 = 10$ kΩ, $R_2 = 10$ kΩ, $R_3 = 10$ kΩ, and $R_4 = 10$ kΩ. Label the **y-axis** as *Voltage Values*. Each point on the graph should represent the value of voltage dropped by the identified resistor. For the Circuit 2 graph, label the **x-axis** with the identifiers: $R_1 = 10$ kΩ, $R_2 = 27$ kΩ, $R_3 = 10$ kΩ, and $R_4 = 47$ kΩ. Again, label the **y-axis** as *Voltage Values*. You may create these graphs by hand on graph paper, or you may use Excel to create the graphs.

57

4. From the measured parameters data in Table 1, perform calculations, as required, to fill in Table 2.

5. After completing Tables 1 and 2, and producing the graphs called for in step 3, answer the Analysis Questions and create a brief Technical Lab Report to complete the project.

Table 1

Identifiers	Circuit 1 Measured Values	Circuit 2 Measured Values
R_1		
R_2		
R_3		
R_4		
R_T		
V_{R_1}		
V_{R_2}		
V_{R_3}		
V_{R_4}		
V_T		
I_T		

Table 2

Identifiers	Calculated P Values for Circuit 1	Calculated P Values for Circuit 2
P_{R_1}		
P_{R_2}		
P_{R_3}		
P_{R_4}		
P_T		

Analysis Questions

NOTE Answers to these Analysis Questions should be clearly numbered and documented on separate sheets of paper with your name and the date at the top of each page. These answer sheets are to be turned in with the rest of the project documentation, as appropriate.

1. From your collected data, is it obvious that the current is the same through all of the components in series? Explain your observation.

2. From your collected data, was each resistor's voltage drop in proportion to its resistance value relative to the other resistors' values? Relative to its portion of total resistance? Give a sample of the parameter measurements that prompted this response.

3. In Circuit 2, which resistor dissipated the most power? Which resistor(s) dissipated the least power? Explain why this is true.

4. In both Circuit 1 and Circuit 2, did the measured resistances precisely match the color-coded resistor values? If not, why not?

5. Did your measured values for R_T in either circuit precisely match the R_T you would have gotten by simply adding the color-coded resistor values?

6. Look at the color-coded tolerances for each resistor and determine what the range of total resistance values could be for each circuit. Determine whether or not your measured values were in the acceptable range.

    Minimum R_T possible if all resistors were *below* color-coded value by the amount of their tolerances: Circuit 1 minimum possible R_T _____ kΩ. Circuit 2 minimum possible R_T _____ kΩ.

    Maximum R_T possible if all resistors were *above* color-coded value by the amount of their tolerances: Circuit 1 maximum possible R_T _____ kΩ. Circuit 2 maximum possible R_T _____ kΩ.

7. If any one of the resistors in either Circuit 1 or Circuit 2 became open, what would happen to the values of: (a) circuit total current, (b) circuit total power, (c) voltage across the opened resistor, and (d) voltage across the unopened resistors?

8. If any of the resistors in either Circuit 1 or Circuit 2 became shorted, what would happen to the values of: (a) total resistance, (b) total current, and (c) total power provided by the source?

Technical Lab Report

Write a brief technical lab report summarizing the technical facts learned from this project. The report should be organized to provide the following:

1. An introductory paragraph describing the type of circuit being analyzed and the key parameters that will be discussed relating to this circuit.

2. A section describing the most important characteristics of this type of circuit that were shown via the collected data in the tables and graphs.

3. Any special facts or characteristics about this type of circuit that were highlighted in answering the Analysis Questions.

4. A practical example of how the information learned in this project might help you in operating, troubleshooting, error analysis, or adjusting a circuit of this type in your home setting, in your training program setting, or in a job setting in the real world.

5. A summary statement listing the most positive aspects of the project and any parts of the project that were difficult because of equipment problems or unclear instructions. Include areas that might be improved.

Summary
Series Circuits

Name: James Quinn Date: 10-9-14

Complete the following review questions, indicating the appropriate response by placing a check in the box next to the correct answer.

1. The total resistance in a series circuit is equal to
   - ☑ the sum of all the Rs
   - ☐ the largest R minus smallest R
   - ☐ neither of these

2. In a series circuit there is
   - ☑ only one path for current
   - ☐ as many paths as there are components

3. The current in a series circuit is
   - ☑ the same through all parts
   - ☐ different through each component

4. The highest voltage drop in a series circuit appears across
   - ☐ the smallest R
   - ☑ the highest R
   - ☐ the average value R

5. If R_1 is ten times larger in value than R_2,
   - ☑ V_{R_1} = ten times V_{R_2}
   - ☐ V_{R_1} = one-tenth V_{R_2}
   - ☐ neither answer is correct

6. In a series circuit, the smallest R will dissipate
   - ☐ the most power
   - ☐ no power
   - ☑ the least power
   - ☐ none of these

7. In a series circuit, the applied voltage equals
   - ☐ the largest V drop minus the smallest V drop
   - ☑ the sum of all V drops

8. In a series circuit, the percentage of the circuit applied voltage dropped across any given resistor is directly related and dependent upon the ratio of the given R to
   - ☐ R total
   - ☐ I total
   - ☑ V total
   - ☐ P total

9. In a series circuit, if part of the circuit becomes open, then
   a. R_T will
      - ☑ increase
      - ☐ decrease
      - ☐ remain the same
   b. V_A will
      - ☐ increase
      - ☑ decrease
      - ☑ remain the same

c. I_T will
- ☐ increase
- ☒ decrease
- ☐ remain the same

d. The voltage across the opened portion of the circuit will
- ☐ increase
- ☒ decrease
- ☐ remain the same

e. The voltage drops across the unopened elements will
- ☐ increase
- ☒ decrease
- ☐ remain the same

10. In a series circuit, if part of the circuit becomes shorted, then

a. R_T will
- ☐ increase
- ☒ decrease
- ☐ remain the same

b. V_A will
- ☐ increase
- ☐ decrease
- ☒ remain the same

c. I_T will
- ☒ increase
- ☐ decrease
- ☐ remain the same

d. The voltage across the shorted portion of the circuit will
- ☒ increase
- ☐ decrease
- ☐ remain the same

e. The voltage drops across the other components in the circuit will
- ☒ increase
- ☐ decrease
- ☐ remain the same

Student Log
For Optional Troubleshooting Exercise

Name: _____ Date: _____

NOTE Complete instructions are located on page xiii of the Preface. It is important that you read these instructions carefully before performing any of the troubleshooting exercises.

Students should write down the starting point symptom given to them by the instructor, and then proceed to log each step used in the troubleshooting process. For example, the first step after learning the starting point symptom information is to identify the area to be bracketed as the "area of uncertainty," and should include all circuitry components that "might" cause the trouble producing the symptoms. The next step is to make a decision about the first test, and so on until the exercise is complete and the suspected trouble is determined.

- Starting point symptom

  _____

- Components and circuitry included in "initial brackets" ("area of uncertainty")

  _____

- First test description (what type of test and where?)

  _____

- Components and circuitry *still* included in "bracketed" area after first test

  _____

- Second test description (what type of test and where?)

  _____

- Components and circuitry *still* included in "bracketed" area after second test

  _____

- Third test description (what type of test and where?)

  _____

- Components and circuitry *still* included in "bracketed" area after third test

  _____

- Fourth test description (what type of test and where?)

  _____

- Components and circuitry *still* included in "bracketed" area after fourth test

  _____

- Fifth test description (what type of test and where?)

  _____

NOTE ➤ Keep repeating the testing, bracketing, and logging procedure until you are quite sure you have found the trouble. Call the instructor to check your work; then, replace the identified faulty component to see if the circuit operates properly.

- Suspected trouble

  _____

- Trouble verified by instructor and by circuit operation: ____ Yes ____ No

PARALLEL CIRCUITS

PART 4

Objectives

You will connect several dc resistive parallel circuits and make measurements and observations regarding their important electrical characteristics.

In completing these projects, you will connect circuits, make measurements, perform calculations, draw conclusions, and be able to answer questions about the following items related to parallel circuits:

- Total resistance
- Voltage distribution
- Current distribution
- Power dissipation(s)
- Effects of opens
- Effects of shorts
- Application of Ohm's law
- Application of Kirchhoff's voltage law
- Application of Kirchhoff's current law

Project/Topic Correlation Information

PROJECT		TEXT CHAPTER	SECTION	RELATED TEXT TOPIC(S)
15	Equivalent Resistance in Parallel Circuits	5	5-5 5-6	Resistance in Parallel Circuits
16	Current in Parallel Circuits	5	5-3	Current in Parallel Circuits
17	Voltage in Parallel Circuits	5	5-1 5-2	Voltage in Parallel Circuits
18	Power Distribution in Parallel Circuits	5	5-7	Power in Parallel Circuits
19	Effects of an Open in Parallel Circuits	5	5-8	Effects of Opens in Parallel Circuits and Troubleshooting Hints
20	Effects of a Short in Parallel Circuits	5	5-9	Effects of Shorts in Parallel Circuits and Troubleshooting Hints

Parallel Circuits
Equivalent Resistance in Parallel Circuits

Name: James Quinn   Date: 10-7-14

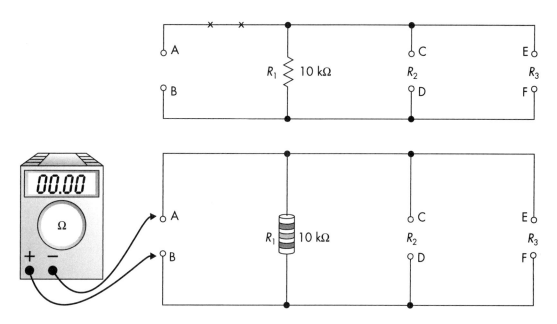

FIGURE 15-1 Standard schematic diagram (top) and pictorial diagram (bottom)

PROJECT PURPOSE To verify that the equivalent resistance of parallel resistances is less than the least value resistance in parallel. To provide further practice in using parallel resistance formulas and confirmation of these through circuit measurements.

PARTS NEEDED
- ☐ DMM
- ☐ CIS
- ☐ Resistors
  10 kΩ (3)   100 kΩ
  47 kΩ

SAFETY HINTS DO NOT use a power supply in the first six (6) steps of this project!

67

68 PART 4: Parallel Circuits

PROCEDURE

1. Connect the initial circuit as shown in Figure 15-1.

2. Measure R_e at points A and B.

   ⚠ OBSERVATION R_e = __10,000__ Ω. *10 kΩ*

3. Use the popular product-over-the-sum formula and calculate R_e if a 10-kΩ R were inserted between points C and D on the trainer.

   $R_e = R_1 \times R_2 \div (R_1 + R_2)$

   ⚠ OBSERVATION R_e calculated = __5,000__ Ω. *5 kΩ*

   ⚠ CONCLUSION When two resistors of equal value are connected in parallel, R_e is equal to one-half the R of (*one, both*) __one__ branch(es).

   After calculating R_e, measure it with a DMM at points A and B with a second 10-kΩ R inserted at points C and D.

   ⚠ OBSERVATION R_e measured = __5 k__ Ω.

4. Assume a third 10-kΩ R is to be added at points E and F and calculate R_e. Use the reciprocal method, or $R_e = R_e{}^1 \times R_3 \div (R_e{}^1 + R_3)$.

   ⚠ OBSERVATION R_e calculated = __3.3 k__ Ω.

   ⚠ CONCLUSION When three resistors of equal value are connected in parallel, R_e is equal to (*1/2, 1/3, 1/4*) __1/3__ the R of one branch.

   After calculating, connect a third 10-kΩ R between points E and F and measure R_e to verify your calculations.

   ⚠ OBSERVATION R_e measured = __3.3__ Ω.

   ⚠ CONCLUSION In parallel circuits, the total resistance of the circuit is always less than the (*highest, lowest*) __lowest__ branch R. If two or more *unequal* Rs are in parallel, can we divide one branch R by the number of branches to find R_e? __no yes__

5. Change R_2 to a 47-kΩ R and R_3 to a 100-kΩ R. Use the assumed voltage method of solving for R_e as follows:
   a. Assume ㊼ volts applied.
   b. Solve for each branch I.
   c. Calculate I_T.
   d. Find R_e by Ohm's law.

   $(R_e = \dfrac{V_T}{I_T})$

⚠ OBSERVATION I_{R_1} = __4.7__ mA. I_T = __6.17__ mA.

I_{R_2} = __1__ mA. R_e = __7.6 k__ Ω.

I_{R_3} = __.47__ mA.

After calculating R_e, use the DMM and measure R_e to verify your calculations.

⚠ OBSERVATION R_e measured = __7.7 k__ Ω.

⚠ CONCLUSION   The assumed voltage method of finding R_e is sometimes easier to use than the product-over-the-sum method, especially when several branches of unequal values are involved. Many times, if an appropriate value of V applied is assumed, one can solve for the branch currents in his/her head, then add the branch currents easily to solve for (R_T, I_T) __I_T__. Then all that remains is a simple (*multiplication, division*) __division__ problem to find R_e.

NOTE ➤ Using a calculator, the reciprocal method is also very easy!

PART 4: Parallel Circuits

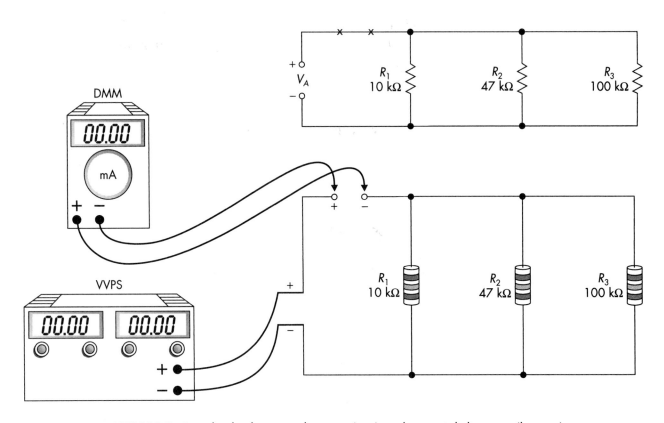

FIGURE 15-2 Standard schematic diagram (top) and pictorial diagram (bottom)

PROJECT PURPOSE To provide further practice in using parallel resistance formulas and confirmation of these through circuit measurements.

PARTS NEEDED
- DMM
- VVPS (dc)
- CIS
- Resistors
  - 10 kΩ   one R to be
  - 47 kΩ   determined by
  - 100 kΩ student calculation

 SAFETY HINTS Be sure power is **OFF** when making circuit changes and/or when connecting meters **AND** when measuring resistance.

PROCEDURE

$\frac{7.6}{10k}$

6. With $R_1 = 10$ kΩ, $R_2 = 47$ kΩ, and $R_3 = 100$ kΩ as shown in Figure 15-2, what will the parameters (electrical circuit values) be if we apply 7.6 volts? (Use 7.6 as the assumed voltage and calculate as before.)

OBSERVATION
I_{R_1} = .7 mA. I_T = .93 mA.
I_{R_2} = .16 mA. R_e = 7.6 k Ω.
I_{R_3} = .07 mA.

PROJECT 15: Equivalent Resistance in Parallel Circuits

⚠ CONCLUSION Changing V applied does not change (R_T, I_T) __R_T__. The voltage that is "assumed" does make a difference in difficulty of computing the final result (R_e). From the preceding steps, it would seem wise to choose a value of assumed voltage that is easily divided by the resistance values for solving branch currents, since no matter what voltage is assumed, it does not change the actual circuit (I, R) __R__.

7. With the circuit described in step 6 connected, insert the milliammeter to read I_T. Apply 7.6 volts to the circuit and measure I_T. After measuring, REMOVE the power supply and current meter from the circuit. Replace the current meter with a jumper wire.

   ⚠ OBSERVATION $I_T =$ __.93__ mA.

   ⚠ CONCLUSION The assumed voltage method of calculating R_e (*has been*, *has not been*) __has been__ verified.

8. A practical problem that sometimes confronts a technician is what value R must be put in parallel with the existing Rs in order to arrive at a desired equivalent resistance. A simple formula that helps solve this is:

$$R_u = \frac{R_k \times R_e}{R_k - R_e}$$

$R_u = \dfrac{27k \times 17k}{27k - 17k}$ $R_u = 45.9\, k\Omega$

where: R_u is R unknown, R_k is R known, and R_e is the desired resultant equivalent R. Assume you have a 27-kΩ R and want an R_e of 17 kΩ. Use the above formula, solve for R_u, then connect the circuit on the matrix using an available resistor as close to the calculated R_u as possible and measure R_e to verify results.

   ⚠ OBSERVATION $R_u =$ __45.9 kΩ__ Ω.

   ⚠ CONCLUSION The formula for solving for the unknown needed parallel resistance to arrive at a given equivalent resistance seems to work, because the results of the practical circuit were close to the theoretical result. It has been proven again that R_e turns out to be (*less*, *more*) __Less__ than the least resistance branch.

Parallel Circuits
Current in Parallel Circuits

Name: James Quinn Date: 10-9-14

Note:
Solid-line directional current flow arrows indicate electron current flow.

Dotted-line directional current flow arrows indicate conventional current flow.

FIGURE 16-1 Standard schematic diagram (top) and pictorial diagram (bottom)

PROJECT PURPOSE To confirm that current through parallel branches is inverse to each branch's resistance value and that total current equals the sum of all branch currents. To provide practical verification of Kirchhoff's current law.

PARTS NEEDED
- ☐ DMM
- ☐ VVPS (dc)
- ☐ CIS
- ☐ Resistors
  - 10 kΩ   47 kΩ
  - 27 kΩ   100 kΩ

SAFETY HINTS Be sure power is off when making circuit changes and/or when connecting meters.

NOTE ➤ Due to using standard-value resistors, some approximating is called for in this project to simply demonstrate concepts.

74 PART 4: Parallel Circuits

PROCEDURE

1. Connect the initial circuit as shown in Figure 16-1.

2. Replace the appropriate jumper with the current meter to read I_1. Apply 5 volts to the circuit and note I_1. Then move the meters and jumpers as required to measure I_2, I_3, and I_T.

    ⚠ OBSERVATION $I_1 =$ __.5__ mA. $I_3 =$ __.11__ mA.
    $I_2 =$ __.19__ mA. $I_T =$ __.8__ mA.

    ⚠ CONCLUSION R_2 is roughly __2.7__ times larger in resistance than R_1. The current through R_2 is roughly __2.5__ the current through R_1. R_3 is roughly __4.7__ times larger in R value than R_1 and its current is roughly __16.5__ of the current through R_1. From this we conclude that the current through parallel branches is (directly, inversely) __inversely__ proportional to the branch Rs. Also, we observe from the measured currents that total circuit current equals the (product, sum) __sum__ of the __total__ currents.

3. Change R_2 from 27 kΩ to 100 kΩ. Measure and record all the circuit currents one at a time. Make V applied 5 volts in each case.

    ⚠ OBSERVATION $I_1 =$ __.5__ mA. $I_3 =$ __.11__ mA.
    $I_2 =$ __.05__ mA. $I_T =$ __.66__ mA.

    ⚠ CONCLUSION   When R_2 was changed from 27 kΩ to 100 kΩ, keeping the same V_A, did the current through R_1 or R_3 change? __no__. Did I_T change? __yes__. If so, was the change in I_T the same as the change in I_2? __yes__. This again proves that I_T is simply the sum of all the __total__ currents. Since R_2 is ten times larger than R_1, its current should be __1/10__ of I_1. Is it? __yes__.

4. Refer again to the circuit diagram, and make the appropriate statements in the "Conclusion" section that will verify Kirchhoff's current law, which states that the current away from any point in a circuit must equal the current to that point.

    ⚠ CONCLUSION   Does the value of electron flow current going away from Point A equal the value of current coming to Point A? _____.
    Does the value of electron flow current going away from Point B equal the value of current coming to Point B? _____.
    Does the value of electron flow current going away from Point C equal the value of current coming to Point C? _____.
    Does the value of electron flow current going away from Point D equal the value of current coming to Point D? _____.

Parallel Circuits
Voltage in Parallel Circuits

PROJECT 17

Name: James Quinn   Date: 10-16-18

FIGURE 17-1 Standard schematic diagram (top) and pictorial diagram (bottom)

PROJECT PURPOSE To demonstrate that the voltage across parallel branches is equal and that, in a simple parallel circuit, each branch voltage equals the circuit applied voltage.

PARTS NEEDED
- ☐ DMM
- ☐ VVPS (dc)
- ☐ CIS
- ☐ Resistors
  10 kΩ (4)   100 kΩ

SAFETY HINTS Be sure power is off when making circuit changes and/or when connecting meters.

PROCEDURE

1. Connect the initial circuit as shown in Figure 17-1.

2. Apply 10 volts V_A to the circuit and measure all the circuit voltages. Record in the "Observation" section.

   OBSERVATION $V_A =$ __10__ V. $V_{R_2} =$ __10__ V.
   $V_{R_1} =$ __10__ V. $V_{R_3} =$ __10__ V.

   CONCLUSION We can conclude from the measurements that the voltage across all the resistors in parallel is the __same__. Also, in a simple, purely parallel circuit configuration, the voltage across each branch equals V __T__.

3. Change the value of R_2 from 10 kΩ to 100 kΩ and measure all the circuit voltages. Record in the "Observation" section.

   OBSERVATION $V_A =$ __10__ V. $V_{R_2} =$ __10__ V.
   $V_{R_1} =$ __10__ V. $V_{R_3} =$ __10__ V.

   CONCLUSION Did all the voltages measure the same as before? __yes__. This indicates that changing the value of branch Rs in a parallel circuit (*does, does not*) __does not__ alter the branch voltages. As a matter of fact, the branch voltages are not separate voltages, but really all the __same__ V. Does changing the value of a branch R change the circuit total current? __yes__. Does it change the current through the unchanged R branch(es)? __no__. Does it change the current through the changed R branch? __yes__.

4. Add a fourth branch to the parallel circuit by inserting a 10-kΩ resistor between points A and B. Measure the circuit voltages with 10 volts applied.

   OBSERVATION $V_A =$ __10__ V. $V_{R_3} =$ __10__ V.
   $V_{R_1} =$ __10__ V. $V_{R_4} =$ __10__ V.
   $V_{R_2} =$ __10__ V.

   CONCLUSION From the results of this step, we conclude that adding additional branches to a parallel circuit (*does, does not*) __does not__ alter the voltage across the parallel branches. (This assumes that the source of voltage is capable of meeting the current requirements of the circuit without its output voltage being altered.)

Parallel Circuits
Power Distribution in Parallel Circuits

Name: James Quinn   Date: 10-16-14

Note:
Solid-line directional current flow arrows indicate electron current flow.

Dotted-line directional current flow arrows indicate conventional current flow.

FIGURE 18-1 Standard schematic diagram (top) and pictorial diagram (bottom)

PROJECT PURPOSE To illustrate that in parallel circuits power dissipation by any given branch is inverse to that branch's resistance value. To also demonstrate that total power in parallel circuits equals the sum of the branch power dissipations.

PARTS NEEDED
- ☐ DMM
- ☐ VVPS (dc)
- ☐ CIS
- ☐ Resistors
  10 kΩ (2)   47 kΩ
  27 kΩ

SAFETY HINTS Be sure power is off when making circuit changes and/or when connecting meters.

PROCEDURE

1. Connect the initial circuit as shown in Figure 18-1.

2. Apply 10 volts to the circuit, replace the appropriate "branch jumpers" one at a time with the milliammeter in order to measure each of the branch currents (I_1, I_2, and I_3), and record the results in the Observation section.

   OBSERVATION $I_1 = $ __1__ mA. $I_3 = $ __.21__ mA.

   $I_2 = $ __.37__ mA.

   CONCLUSION Using the law that total current in a parallel circuit is equal to the (*product*, *sum*) __Sum__ of the __resistor__ currents, the total current for the circuit must be __1.58__ mA. Since the total power dissipated in any resistive circuit can be calculated as V_T times __I_T__, then the total power dissipated by this circuit and being supplied by the source is __15.8__ mW.

3. Use the appropriate power formula(s) and calculate the individual power dissipations of R_1, R_2, and R_3. Record your answers.

   OBSERVATION $P_{R_1} = $ __10 mW__ mW. $P_{R_3} = $ __2.1__ mW.

   $P_{R_2} = $ __3.7__ mW.

   CONCLUSION Does the sum of $P_{R_1} + P_{R_2} + P_{R_3}$ equal P_T? __yes__. Which resistor dissipates the most power? __R1__. This resistor is the (*largest*, *smallest*) value R in the circuit. Which resistor dissipated the least power? __R3__. This resistor is the (*largest*, *smallest*) __largest__ value R in the circuit. From this we conclude that in a parallel circuit the smaller the branch R, the (*lesser*, *greater*) __greater__ power it will dissipate, because all *V*s are the same, and the *I*s are inversely proportional to the branch *R*s. Therefore, the $V \times I$ product will be greater if the resistor is of (*low*, *high*) __low__ value.

4. Change R_2 from a 27-kΩ to a 10-kΩ resistor. Measure the branch currents and calculate the individual power dissipations and the P_T with 10 volts applied.

   OBSERVATION $I_1 = $ _____ mA. $P_{R_2} = $ _____ mW.

   $I_2 = $ _____ mA. $P_{R_3} = $ _____ mW.

   $I_3 = $ _____ mA. $P_T = $ _____ mW.

   $P_{R_1} = $ _____ mW.

   CONCLUSION Changing the resistance of one branch in a parallel circuit will cause the power dissipated by that branch to change and the _____ power to change. But the power dissipated by the other branches will remain the same as long as their resistance values are not changed.

Parallel Circuits
Effects of an Open in Parallel Circuits

PROJECT 19

Name: James Quinn   Date: 10-23-14

Note:
Solid-line directional current flow arrows indicate electron current flow.

Dotted-line directional current flow arrows indicate conventional current flow.

FIGURE 19-1 Standard schematic diagram (top) and pictorial diagram (bottom)

PROJECT PURPOSE To provide experience through circuit measurements and observations regarding what happens to circuit parameters when an open occurs in a parallel circuit. To verify that total circuit current will decrease by the amount of current that was passing through the defective branch, prior to its opening.

PARTS NEEDED
- [ ] DMM
- [ ] VVPS (dc)
- [ ] CIS
- [ ] Resistors 47 kΩ   100 kΩ (2)

SAFETY HINTS. Be sure power is off when making circuit changes and/or when connecting meters.

PROCEDURE

1. Connect the initial circuit as shown in Figure 19-1.

2. Apply 20 volts to the circuit and note the measured total current. Use Ohm's law and calculate R_T and the branch currents.

    OBSERVATION $I_T =$ _____ mA. $I_2 =$ _____ mA.

    $R_T =$ _____ Ω. $I_3 =$ _____ mA.

    $I_1 =$ _____ mA.

    CONCLUSION R_T is less than the (*highest, lowest*) _____ resistance branch. I_T is equal to the _____ of the branch currents. The ratios of the branch currents to each other is inverse to the ratios of their _____ to each other.

3. Remove R_1 from the circuit to simulate an open in that branch. Apply 20 volts to the circuit and measure I_T, I_1, I_2, and I_3. Calculate R_T.

    OBSERVATION $I_T =$ _____ mA. $I_2 =$ _____ mA.

    $R_T =$ _____ Ω. $I_3 =$ _____ mA.

    $I_1 =$ _____ mA.

    CONCLUSION   It can be concluded that if any branch of a parallel circuit opens, R_T will (*increase, decrease*) _____; therefore, I_T will (*increase, decrease*) _____ since V remained the same. Did the current through R_2 and R_3 change when R_1 opened? _____. The reason R_T increased was that when R_1 opened there was one less _____ path. The current through the opened branch (*increased, decreased*) _____ to _____ mA; whereas the current through the unopened branches (*increased, decreased, remained the same*) _____. The voltage across all branches (*did, did not*) _____ change when the R_1 branch was opened.

4. Use the proper formula(s) and calculate the branch power dissipations and P_T for the normal circuit condition of step 2.

    OBSERVATION $P_T =$ _____ mW. $P_{R_2} =$ _____ mW.

    $P_{R_1} =$ _____ mW. $P_{R_3} =$ _____ mW.

5. Use the proper formula(s) and calculate the branch power dissipations and P_T for the circuit conditions when R_1 was open in step 3.

⚠ OBSERVATION $P_T =$ _____ mW. $P_{R_2} =$ _____ mW.

$P_{R_1} =$ _____ mW. $P_{R_3} =$ _____ mW.

⚠ CONCLUSION Opening one branch of a parallel circuit (*does, does not*) _____ affect the power dissipated by the other branches. However, total P will (*increase, decrease*) _____ by the amount of power that was being dissipated by the opened branch previous to its opening.

Parallel Circuits
Effects of a Short in Parallel Circuits

PROJECT 20

Name: James Quinn   Date: 10-30-14

FIGURE 20-1 Standard schematic diagram (top) and pictorial diagram (bottom)

PROJECT PURPOSE To provide experience through "POWER-OFF" ohmmeter measurements regarding what happens to parallel circuit parameters, should one or more branches become shorted.

PARTS NEEDED
- ☐ DMM
- ☐ CIS
- ☐ Resistors
  47 kΩ   100 kΩ (2)

SAFETY HINTS DO NOT use a power supply for this project!

83

PART 4: Parallel Circuits

PROCEDURE

1. Connect the initial circuit as shown in Figure 20-1.

   ! CAUTION: Do not use the power supply or milliammeter for this demonstration or project.

2. Use the DMM and measure R_e at points A and B.

   ⚠ OBSERVATION $R_e = $ ~~____~~ Ω. 24 kΩ

3. To simulate R_1 becoming shorted, remove R_1 and replace it with a jumper wire; then measure R_e again.

   ⚠ OBSERVATION $R_e = $ __1.7__ Ω.

   ⚠ CONCLUSION   A short across one branch of a parallel circuit causes R_e to (increase, <u>decrease</u>) _____ effectively to __0__ Ω. If we were using a power supply that could supply infinite current, the current through R_2 would be __0__ mA; through R_3 would be __0__ mA; and through the shorted branch would be __0__ mA. Since $V = I \times R$, what is the IR drop across the shorted branch? ____ V. In parallel, the voltage across all branches in parallel is the _____ voltage; therefore, V_2 and V_3 for the conditions described above would be _____ V.

4. Replace R_1 into the circuit and simulate a short in branch R_2 by replacing R_2 with a jumper wire. Measure R_e again.

   ⚠ OBSERVATION $R_e = $ _____ Ω.

   ⚠ CONCLUSION   A short of *any* branch of a parallel circuit will cause R_e to (*increase, decrease*) _____ and I_T to (*increase, decrease*) _____ (before the power-supply fuse blows). Also, current through the "unshorted" branches would (*increase, decrease*) _____, and current through the "shorted" branch would (*increase, decrease*) _____. The voltage across all branches would (*increase, decrease*) _____. It should be noted that in the strictest sense, if any branch of a parallel circuit is "shorted," all branches are shorted. However, only one branch "contains" the actual shorted element that is "shunting" the whole circuit with the undesired low resistance path, whose resistance approaches (∞, *0*) _____ Ω.

Story Behind the Numbers
Parallel Circuits

Name: _____   Date: _____

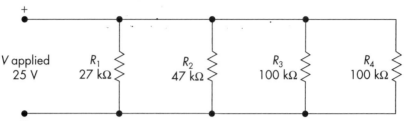

Procedure

NOTE When performing the procedure steps, you have the options of using a calculator or using an Excel spreadsheet program "worksheet" for any required calculations. You may also use Excel for creating tables and for generating graphs.

1. Connect the "initial circuit" as shown above.

2. Assume that all resistors are precisely equal to their color-coded values and that the value of V_A is as shown in the schematic. Calculate R_T. Use Ohm's law and calculate I_T. Use the *current divider rule* formulas and determine the theoretical (calculated) values for the following parameters: I_{R_1}, I_{R_2}, I_{R_3}, I_{R_4}. Record all the calculated results in Table 1, as appropriate.

3. Using a DMM and proper measurement and safety procedures, measure V_A, I_{R_1}, I_{R_2}, I_{R_3}, I_{R_4}, I_T, and R_T. **CAUTION: Remove the power source from the circuit when measuring R_T.** Record these measured values in Table 1, as appropriate.

4. Remove R_3 from the initial circuit to simulate it becoming open. Measure R_T and I_T after R_3 is removed. **Don't forget to remove the power source when measuring R_T.** Record this data in Table 1, as appropriate.

5. Use the *theoretical* values and the *measured* values from Table 1 and determine the various power dissipation values throughout the *initial* circuit and the *modified* circuit. Log your results in Table 2, as appropriate.

6. Use the measured values data for the *Initial Circuit* from Table 1 and create a line graph showing the relationship of branch current versus branch resistance for the given applied voltage. Use the resistor identifiers along the **x-axis.** Let the **y-axis** represent current values.

7. Use the power dissipation values determined for the *Initial Circuit* from the measured values shown in Table 2 to create a line graph of power

dissipation versus branch resistance. Let the **x-axis** represent the resistor identifiers and the **y-axis** show the power dissipation values.

8. After completing Table 1 and Table 2 and producing the line graphs, as indicated, answer the Analysis Questions and create a brief Technical Lab Report to complete the project.

Table 1

Identifiers	Initial Circuit Theoretical (Calculated) Values	Initial Circuit Measured Values	Modified Circuit (R_3 Removed) Measured Values
V_A			—
I_{R_1}			—
I_{R_2}			
I_{R_3}			—
I_{R_4}			
I_T			
R_T			

Table 2

Identifiers	Using Initial Circuit Theoretical (Calculated) Values	Using Initial Circuit Measured Values	Using Modified Circuit (R_3 Removed) Measured Values
P_{R_1}			—
P_{R_2}			—
P_{R_3}			—
P_{R_4}			—
P_T			

Analysis Questions

NOTE Answers to these Analysis Questions should be clearly numbered and documented on separate sheets of paper with your name and the date at the top of each page. These answer sheets are to be turned in with the rest of the project documentation, as appropriate.

1. From your analysis of the collected data, state the relationship of branch currents to branch resistances for *any* given circuit applied voltage.

2. From your analysis of the collected data, state the relationship of branch power dissipations to branch resistances for *any* given circuit applied voltage.

3. What is the approximate ratio of power dissipated by the 47-kΩ resistor to the power dissipated by the 100-kΩ resistor?

4. Describe what happened to total current and total power when R_3 was removed from the circuit. Explain the cause of this change.

5. Using the standard rule regarding power ratings of resistors versus the power they will dissipate in a given circuit, what is the minimum power rating you would use for R_1?

Technical Lab Report

Write a brief technical lab report summarizing the technical facts learned from this project. The report should be organized to provide the following:

1. An introductory paragraph describing the type of circuit being analyzed and the key parameters that will be discussed relating to this circuit.

2. A section describing the most important characteristics of this type of circuit that were shown via the collected data in the tables and graphs.

3. Any special facts or characteristics about this type of circuit that were highlighted in answering the Analysis Questions.

4. A practical example of how the information learned in this project might help you in operating, troubleshooting, error analysis, or adjusting a circuit of this type in your home setting, in your training program setting, or in a job setting in the real world.

5. A summary statement listing the most positive aspects of the project and any parts of the project that were difficult because of equipment problems or unclear instructions. Include areas that might be improved.

Summary
Parallel Circuits

Name: _____   Date: _____

Complete the following review questions, indicating the appropriate response by placing a check in the box next to the correct answer.

1. A greater change of R_e occurs if the resistance that is added in shunt with the existing circuit has a resistance value that is
   - ☐ high
   - ☐ low

2. Adding another resistor in parallel with an existing circuit will cause the circuit's total current to
   - ☐ increase
   - ☐ remain the same
   - ☐ decrease

3. Adding another resistor in parallel with an existing parallel circuit will cause the currents through the original branches to
   - ☐ increase
   - ☐ remain the same
   - ☐ decrease

4. In a simple two-branch parallel circuit consisting of a 2-Ω and a 3-Ω resistor in parallel, three-fifths of the total circuit current will flow through
   - ☐ the 3-Ω R
   - ☐ neither R
   - ☐ the 2-Ω R

5. The total resistance of the circuit in question 4 is
   - ☐ 8 Ω
   - ☐ 2.1 Ω
   - ☐ 2 Ω
   - ☐ 3 Ω
   - ☐ 1.2 Ω
   - ☐ none of these

6. In a circuit consisting of 100-kΩ, 47-kΩ, and 1-kΩ branches, the total circuit resistance must be
   - ☐ more than 1 kΩ
   - ☐ more than 100 kΩ
   - ☐ less than 1 kΩ
   - ☐ more than 47 kΩ

7. The important fact to remember about parallel circuits is that the voltage across all parallel branches is
   - ☐ divided equally
   - ☐ the same
   - ☐ different
   - ☐ none of these

8. When one branch of a parallel circuit opens
   a. V_A will
   - ☐ increase
   - ☐ remain the same
   - ☐ decrease

   b. I_T will
   - ☐ increase
   - ☐ remain the same
   - ☐ decrease

c. The unopened branch currents will
   - ☐ increase
   - ☐ remain the same
   - ☐ decrease
d. The opened branch currents will
   - ☐ increase
   - ☐ remain the same
   - ☐ decrease
e. R_T will
   - ☐ increase
   - ☐ remain the same
   - ☐ decrease

9. If any branch of a parallel circuit "shorts," it will cause
   - ☐ excessive current
   - ☐ excessive resistance
   - ☐ excessive voltage
   - ☐ none of these

10. In a parallel circuit, the resistor that dissipates the most power is
    - ☐ the lowest value R
    - ☐ neither of these
    - ☐ the highest value R

Student Log
For Optional Troubleshooting Exercise

Name: _____ Date: _____

NOTE ➤ Complete instructions are located on page xiii of the Preface. It is important that you read these instructions carefully before performing any troubleshooting exercises.

Students should write down the starting point symptom given to them by the instructor, and then proceed to log each step used in the troubleshooting process. For example, the first step after learning the starting point symptom information is to identify the area to be bracketed as the "area of uncertainty," and should include all circuitry components that "might" cause the trouble producing the symptoms. The next step is to make a decision about the first test, and so on until the exercise is complete and the suspected trouble is determined.

- Starting point symptom

  _____

- Components and circuitry included in "initial brackets" ("area of uncertainty")

  _____

- First test description (what type of test and where?)

  _____

- Components and circuitry *still* included in "bracketed" area after first test

  _____

- Second test description (what type of test and where?)

  _____

- Components and circuitry *still* included in "bracketed" area after second test

  _____

- Third test description (what type of test and where?)

  _____

- Components and circuitry *still* included in "bracketed" area after third test

  _____

- Fourth test description (what type of test and where?)

  _____

- Components and circuitry *still* included in "bracketed" area after fourth test

  _____

- Fifth test description (what type of test and where?)

  _____

NOTE ➤ Keep repeating the testing, bracketing, and logging procedure until you are quite sure you have found the trouble. Call the instructor to check your work; then, replace the identified faulty component to see if the circuit operates properly

- Suspected trouble

  _____

- Trouble verified by instructor and by circuit operation: ____ Yes ____ No

SERIES-PARALLEL CIRCUITS

PART 5

Objectives

You will connect several dc resistive series-parallel circuits and make measurements and observations regarding their important electrical characteristics.

In completing these projects, you will connect circuits, make measurements, perform calculations, draw conclusions, and be able to answer questions about the following items related to series-parallel circuits:

- Total resistance
- Voltage distribution
- Current distribution
- Power dissipation(s)
- Effects of opens
- Effects of shorts
- Application of Ohm's law
- Application of Kirchhoff's voltage law
- Application of Kirchhoff's current law

Project/Topic Correlation Information

PROJECT		TEXT CHAPTER	SECTION	RELATED TEXT TOPIC(S)
21	Total Resistance in Series-Parallel Circuits	6	6-3	Total Resistance in Series-Parallel Circuits
22	Current in Series-Parallel Circuits	6	6-4	Current in Series-Parallel Circuits
23	Voltage Distribution in Series-Parallel Circuits	6	6-5	Voltage in Series-Parallel Circuits
24	Power Distribution in Series-Parallel Circuits	6	6-6	Power in Series-Parallel Circuits
25	Effects of an Open in Series-Parallel Circuits	6	6-7	Effects of Opens in Series-Parallel Circuits and Troubleshooting Hints
26	Effects of a Short in Series-Parallel Circuits	6	6-8	Effects of Shorts in Series-Parallel Circuits and Troubleshooting Hints

Series-Parallel Circuits
Total Resistance in Series-Parallel Circuits

PROJECT 21

Name: James Quinn   Date: 10-16-14

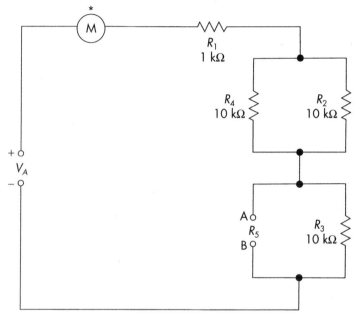

* Use best range to measure currents between 0–1 mA

FIGURE 21-1

NOTE ▶ From this point on, you will notice that there are no pictorial diagrams. You will be expected to wire your circuits and make measurements following the schematic diagram only.

PROJECT PURPOSE To use and verify series-parallel circuit resistance analysis techniques by circuit measurements and/or related calculations.

PARTS NEEDED
- ☐ DMM
- ☐ VVPS (dc)
- ☐ CIS
- ☐ Resistors
  1 kΩ   10 kΩ (4)

SAFETY HINTS From this point on, we will not repeat the standard safety hints you have been seeing thus far but will reserve this box for special safety hints, as appropriate.

NOTE ▶ The term R_T generally refers to total circuit resistance, whereas the term R_e (R equivalent) may refer to the total resultant resistance of specific parallel resistors in just a portion of the circuit. If the total circuit is a purely parallel circuit, the term R_e is the same as R_T, for that particular case.

95

96 PART 5: Series-Parallel Circuits

PROCEDURE

1. Connect the initial circuit as shown in Figure 21-1.

2. Calculate the circuit R_T from the indicated resistance values as follows: Add $R_1 + R_e$ of R_2 and R_4 (in parallel) + R_3.

   OBSERVATION $R_T = $ __11 kΩ__ Ω.

   CONCLUSION   Is the total resistance of this circuit less than the least resistor? __No__. Is the total resistance of this circuit equal to the sum of all the individual resistors? __No__.

3. Apply just enough voltage to the circuit to obtain 1 mA of current and measure V_A, V_{R_1}, V_{R_2}, V_{R_3}, and V_{R_4}.

   OBSERVATION
   $V_A = $ __11 V__ V. $V_{R_3} = $ __10 V__ V.
   $V_{R_1} = $ __1 V__ V. $V_{R_4} = $ __10 V__ V.
   $V_{R_2} = $ __10 V__ V.

   VA = 1m · 11k
   VA = V
   V_{R_1} = 1m · 1

   CONCLUSION   Since there is __1__ mA of total circuit current and V_A is __11__ V, then by Ohm's law, R_T must be __11K__ Ω. Which resistors in this circuit are in parallel with one another? __R4__ and __R2__. Is the voltage across these two Rs the same? __yes__. Which resistor(s) are in series with the main line (carry I_T)? __R1 and R3__.

4. DISCONNECT the *power supply and milliammeter* from the circuit and measure R_T (at appropriate points in the circuit) with an ohmmeter.

   OBSERVATION $R_T = $ __11 k__ Ω.

   CONCLUSION   It can be concluded that in series-parallel circuits the total resistance is equal to the sum of all the components in (series, parallel) __series__ with the main line (components through which I_T passes) plus the equivalent resistance of parallel elements, whose R_e is effectively in (series, parallel) __series__ with the main line components.

5. Using the logic just discussed, calculate what the circuit R_T would be if a 10-kΩ resistor were added to the circuit between points A and B. After calculating R_T, measure it with a DMM to verify your thinking (meter and power supply still not connected).

   OBSERVATION $R_{calc} = $ __11 k__ Ω. $R_{meas} = $ __14k__ Ω.

   CONCLUSION R_T was calculated by adding the resistance of R __5__ + the R_e of R __3__ and R __4__ + the R_e of R __2__ and R __5__. Would R_T change if the jumper were removed that runs between the top of R_3 and the top of R_5? __yes__.

Optional Steps

6. Assume that the 10-kΩ resistor is left in the circuit between points A and B. Also assume that the circuit is to be rewired so that all four 10-kΩ resistors will be in parallel with each other, and that combination is to still be in series with R_1 (the 1-kΩ resistor). Predict the total resistance for the conditions just described.

   ⚠ OBSERVATION $R_{predicted} = $ _____ kΩ.

   ⚠ CONCLUSION   Is the predicted resistance higher or lower than that measured in step 5? _____.

   Why? _____

   _____.

7. Actually rewire the circuit so that the four 10-kΩ resistors are in parallel with each other, and in series with the 1-kΩ resistor; then, *without any power applied to the circuit,* measure and record the circuit total resistance.

   ⚠ OBSERVATION $R_{meas} = $ _____ kΩ.

   ⚠ CONCLUSION   Did the measured value reasonably agree with your predicted value in the preceding step? _____. If a 1-kΩ resistor had been placed across points A and B rather than the 10-kΩ resistor, would total resistance have been higher or lower? _____. Logic and the measurements of this project seem to indicate that a low R value placed in parallel with an existing S-P circuit will affect the circuit's total resistance (*more, less*) _____ than a larger R value being placed in parallel with the existing circuit.

8. Actually replace R_5 (across points A and B) with a 1-kΩ resistor; then, measure total resistance to see if it verifies your conclusions.

   ⚠ OBSERVATION $R_{meas} = $ _____ kΩ.

   ⚠ CONCLUSION   Are the conclusions verified? _____.

Series-Parallel Circuits
Current in Series-Parallel Circuits

Name: _____   Date: _____

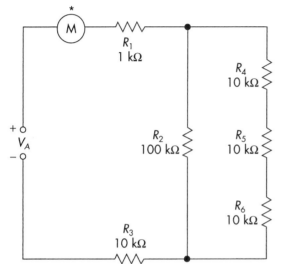

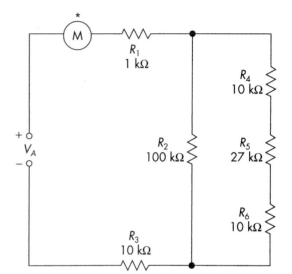

* Use best range to measure currents between 0–1 mA

FIGURE 22-1

* Use best range to measure currents between 0–1 mA

FIGURE 22-2

PROJECT PURPOSE To confirm the characteristics of current division and summation through a series-parallel circuit, using measurements, observations, and both series and parallel circuit analysis techniques. To observe that a component's location, as well as its value, has significance regarding its electrical parameters when it is in a series-parallel circuit.

PARTS NEEDED
- ☐ DMM
- ☐ VVPS (dc)
- ☐ CIS
- ☐ Resistors
  - 1 kΩ (2)   27 kΩ
  - 10 kΩ (4)   100 kΩ

PROCEDURE

1. Connect the initial circuit as shown in Figure 22-1.

2. Apply just enough voltage to the circuit to obtain 0.5 mA of I_T. Measure all the individual voltage drops and calculate the current through each component as required to fill in the blanks in the "Observation" section.

⚠ **OBSERVATION**

V_{R_1} = _____ V. V_{R_4} = _____ V.

V_{R_2} = _____ V. V_{R_5} = _____ V.

V_{R_3} = _____ V. V_{R_6} = _____ V.

99

100 PART 5: Series-Parallel Circuits

$V_A =$ _____ V. $I_{R_4} =$ _____ mA.
$I_{R_1} =$ _____ mA. $I_{R_5} =$ _____ mA.
$I_{R_2} =$ _____ mA. $I_{R_6} =$ _____ mA.
$I_{R_3} =$ _____ mA.

⚠ CONCLUSION   The components that are in series with the main line and through which total current flows are: R _____ and R _____. The components that are in series with each other, but not the main line are: _____, _____, and _____. What value of resistance is in parallel with R_2? _____ Ω. Does the current divide through the parallel branches inverse to the resistance relationships? _____. It should be noted that the current and voltage distribution in a series-parallel circuit for any given component is dependent upon its location in the circuit. However, the series "sections" of the circuit can be analyzed by the same rules as used in _____ circuits, and the parallel sections of the circuit can be analyzed by the rules for _____ circuits. To finalize the circuit analysis, you then combine the results of the sectional analysis.

3. Analyze the circuit and predict what will happen to all the currents in the circuit if R_5 is changed to a 27-kΩ resistor. Will they *increase, decrease,* or *remain the same*? Note your predictions in the "Observation" section.

⚠ OBSERVATION
I_{R_1} will _____. I_{R_5} will _____.
I_{R_2} will _____. I_{R_6} will _____.
I_{R_3} will _____. I_T will _____.
I_{R_4} will _____.

⚠ CONCLUSION   Increasing the value of any resistor in a series-parallel circuit (or any other circuit) will cause I_T to (*increase, decrease*) _____. It is possible for current to increase through a component in parallel with the changed value R because a larger percentage of I_T (even though I_T has decreased some) will pass through the unchanged branch.

4. To verify your predictions of the previous step 3 in Figure 22-1, change R_5 from a 10-kΩ resistor to a 27-kΩ resistor as shown in the diagram in Figure 22-2. Measure all the circuit voltages with 17 volts applied to the circuit. Calculate the current through each component using Ohm's law. Fill in the blanks as required in the "Observation" section. Compare these results with those of the previous step 2 in Figure 22-1 to see if the current(s) increased or decreased.

⚠ OBSERVATION
$V_{R_1} =$ _____ V. $I_{R_1} =$ _____ mA.
$V_{R_2} =$ _____ V. $I_{R_2} =$ _____ mA.
$V_{R_3} =$ _____ V. $I_{R_3} =$ _____ mA.
$V_{R_4} =$ _____ V. $I_{R_4} =$ _____ mA.
$V_{R_5} =$ _____ V. $I_{R_5} =$ _____ mA.
$V_{R_6} =$ _____ V. $I_{R_6} =$ _____ mA.
$V_A =$ _____ V. $I_T =$ _____ mA.

⚠ CONCLUSION   Do the measurements agree with the predictions? _____. Another way of analyzing why I_{R_2} increased even though I_T decreased is to consider the circuit to be a "simple" series circuit of R_1 in series with the R_e of R_2 in parallel with R_4, R_5, R_6, and R_3. Thus, the "equivalent" series circuit is $R_1 + R_e + R_3$. Because R_5 was increased from 10 kΩ to 27 kΩ, R_e will increase. Now R_e is a bigger percentage of R_T, and thus more of V_A will be dropped across R_e and hence R_2. Since R_2 has not itself changed value, yet V_{R_2} is higher, the current through R_2 must also be higher.

5. Remove the current meter. Change R_5 from 27 kΩ to 1 kΩ (which is lower than the original circuit's 10 kΩ). As time permits, predict the circuit parameter "trends" (as opposed to specific values). After predicting, make any measurements required to verify or correct the predictions.

   ⚠ OBSERVATION   The currents that will increase are: _____

   _____.

   The currents that will decrease are: _____

   _____.

   ⚠ CONCLUSION   Decreasing the value of any resistor in a series-parallel circuit (or any other circuit) will cause I_T to (*increase, decrease*) _____. The current through R _____ decreased because with a higher I_T the $I \times R$ drops increased across R _____ and R _____, which means less of V_A was left to be dropped by R _____. Since this resistor has not changed value and it dropped less V, then the current through it must have (*increased, decreased*) _____.

Optional Step

6. Use the circuit of Figure 22-2; however, change R_1 from 1 kΩ to 12 kΩ. Use 17 volts applied voltage. Measure and calculate, as appropriate, to fill in blanks and draw conclusions.

   ⚠ OBSERVATION $I_T =$ _____ mA. $I_{R_4} =$ _____ mA.

   $I_{R_2} =$ _____ mA.

   ⚠ CONCLUSION   Does R_1 carry total current? _____. Did the circuit total current change when R_1 was changed? _____. Did the value of current through R_2 change? _____. Did the percentage of total current through the R_2 branch change when R_1 was changed? _____. (Be careful when thinking about this question!) Did the percentage of total current carried by the branch with R_4 in it change when R_1 was changed? _____. This indicates that changing a series component that carries total current in a series-parallel circuit will cause a change in total circuit current, but (*will, will not*) _____ affect the division of current through parallel branches in the circuit.

Series-Parallel Circuits
Voltage Distribution in Series-Parallel Circuits

PROJECT 23

Name: _____ Date: _____

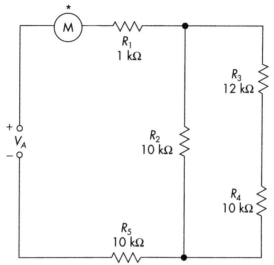

* Use best range to measure currents between 0–1 mA

FIGURE 23-1

PROJECT PURPOSE To observe voltage distribution characteristics in a series-parallel circuit, using measurements and observations. To prove that in a series-parallel circuit, a component's location, as well as its value, affects its voltage drop.

PARTS NEEDED
- ☐ DMM
- ☐ VVPS (dc)
- ☐ CIS
- ☐ Resistors
  - 1 kΩ   12 kΩ
  - 10 kΩ (3)

PROCEDURE

1. Connect the initial circuit as shown in Figure 23-1.

2. Assume an I_T of 0.5 mA and calculate all the circuit voltages. Note results of calculations.

   NOTE ➤ Solve for the R_e of branch R_2 in parallel with branch ($R_3 + R_4$) so you can use the current-divider rule(s) to find currents through these particular branch resistors. This will then allow you to find their individual resistor voltage drops, using the appropriate $I \times R$ values.

103

⚠ **OBSERVATION** $V_{R_1} =$ _____ V. $V_{R_4} =$ _____ V.

$V_{R_2} =$ _____ V. $V_{R_5} =$ _____ V.

$V_{R_3} =$ _____ V.

⚠ **CONCLUSION** According to Kirchhoff's voltage law, the applied voltage to obtain these results would be _____ V.

3. Apply just enough V_A to obtain 0.5 mA of I_T and measure all the circuit voltages.

⚠ **OBSERVATION** $V_{R_1} =$ _____ V. $V_{R_4} =$ _____ V.

$V_{R_2} =$ _____ V. $V_{R_5} =$ _____ V.

$V_{R_3} =$ _____ V. $V_A =$ _____ V.

⚠ **CONCLUSION** Did the measurements verify the calculations of step 2? _____. Did the largest value R drop the most voltage? _____. Because of its location in the circuit, only a small portion of the total _____ passes through this component. The important consideration when analyzing voltage distribution in a series-parallel circuit is the electrical location of each component being considered. I_T only passes through those components that are in series with the (*largest branch, main line*) _____ _____ or source. Parallel sections of the circuit must be analyzed using parallel circuit rules. However, the "net R_e" of a parallel section can be considered as being in series with (*largest branch, main line*) _____ components.

4. Predict what will happen to V_{R_2} if R_3 is replaced with a jumper wire short. After making the prediction, make the suggested circuit change and note the results.

⚠ **OBSERVATION** V_{R_2} will _____. V_{R_2} did _____.

⚠ **CONCLUSION** Decreasing the R in the parallel branch causes total current to (*increase, decrease*) _____, which causes the components that are in series with the main line $I \times R$ drops to (*increase, decrease*) _____. Thus, less of V_A is dropped across R_2.

Series-Parallel Circuits
Power Distribution in Series-Parallel Circuits

PROJECT 24

Name: _____   Date: _____

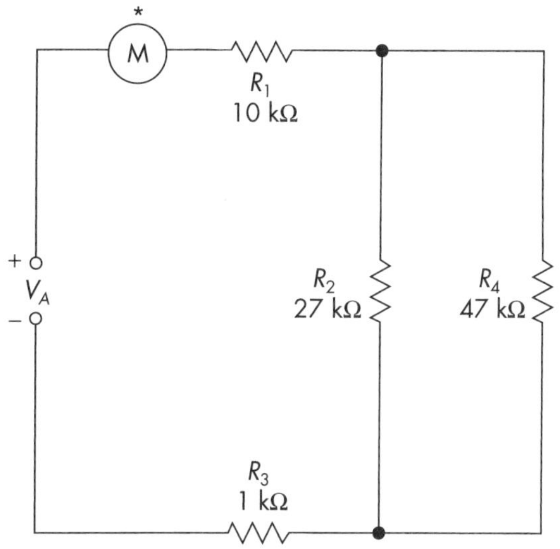

FIGURE 24-1

* Use best range to measure currents between 0–1 mA

PROJECT PURPOSE To connect a series-parallel circuit, make measurements and perform calculations that illustrate the power distribution characteristics of this type of circuit. To illustrate that it is the "electrical location" of a given component, not the physical location of a component, which influences its electrical parameters.

PARTS NEEDED
- ☐ DMM
- ☐ VVPS (dc)
- ☐ CIS
- ☐ Resistors
  - 1 kΩ   27 kΩ
  - 10 kΩ   47 kΩ

PROCEDURE

1. Connect the initial circuit as shown in Figure 24-1.

2. Apply just enough V_A to obtain 0.5 mA of current through the meter.

   ⚠ OBSERVATION $V_A =$ _____ V.
                               (approximately)

   ⚠ CONCLUSION R_T must equal V_T/I_T or _____ kΩ.

3. Measure all voltage drops and calculate the power dissipated by each component.

OBSERVATION

$V_{R_1} =$ _____ V. $P_{R_2} =$ _____ mW.

$V_{R_2} =$ _____ V. $P_{R_3} =$ _____ mW.

$V_{R_3} =$ _____ V. $P_{R_4} =$ _____ mW.

$V_{R_4} =$ _____ V. $P_T =$ _____ mW.

$P_{R_1} =$ _____ mW.

CONCLUSION

The power dissipated by each component in this circuit is directly proportional to its _____ drop and the _____ through it. This statement is true for any circuit. For the parallel sections of the circuit, the larger the R, the (*more, less*) _____ the power dissipated by that component. For the components in series with the main line, the larger the R, the (*more, less*) _____ the power dissipated by that component. Which component in the circuit dissipated the most power? _____. Notice this is not the largest R in the circuit, as it would have to be in a simple series circuit if it dissipated the most power. Also, it is not the smallest R, as it would have to be in a simple parallel circuit if it dissipated the most power. The key to the power distribution for series-parallel circuits is again the electrical (*size, location*) _____ of the component being considered in the circuit.

4. Swap the positions of R_2 and R_4. Make measurements and determine if the powers dissipated by each of the Rs have changed. (Measure voltages and compare to step 3 results.)

OBSERVATION Results are: (*same, different*) _____.

CONCLUSION It is not the "physical" location that is important but rather the _____ location of components that determines the distribution of power (and other parameters) in a series-parallel circuit.

Series-Parallel Circuits
Effects of an Open in Series-Parallel Circuits

Name: _____ Date: _____

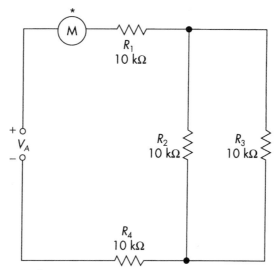

* Use best range to measure currents between 0–1 mA

FIGURE 25-1

PROJECT PURPOSE To demonstrate, through measurements and calculations, the various effects on electrical parameters of an open occurring in a series portion and in a parallel portion of a series-parallel circuit. To note the effect on the overall circuit total current when an open occurs.

PARTS NEEDED ☐ DMM ☐ Resistors
 ☐ VVPS (dc) 10 kΩ (4)
 ☐ CIS

PROCEDURE

1. Connect the initial circuit as shown in Figure 25-1.

2. Increase V applied until 1 mA of current is indicated by the milliammeter. Measure and note the circuit voltages.

   ⚠ OBSERVATION $V_A =$ _____ V. $V_{R_3} =$ _____ V.
   $V_{R_1} =$ _____ V. $V_{R_4} =$ _____ V.
   $V_{R_2} =$ _____ V.

107

3. Calculate the current through each component using Ohm's law.

   OBSERVATION I_{R_1} = _____ mA. I_{R_3} = _____ mA.

   I_{R_2} = _____ mA. I_{R_4} = _____ mA.

   CONCLUSION   Total current flows through R _____ and R _____. The current divides equally between R _____ and R _____ as they are _____ resistances and are in _____, thus they have the same voltage drop.

4. Remove R_1 from the circuit to simulate R_1 "opening." Measure the circuit voltages.

   OBSERVATION V_A = _____ V. V_{R_3} = _____ V.

   V_{R_1} = _____ V. V_{R_4} = _____ V.
   (across the open)

   V_{R_2} = _____ V.

   CONCLUSION   Opening a component in series with the source caused I_T to (*increase, decrease*) _____ to _____ mA. The voltage drops across all the good components decreased to _____ V. The potential difference across the open became equal to V _____. What electrical law does this verify? _____ voltage law.

5. Replace R_1 into the circuit. Now remove R_3 to simulate it "opening." Measure the circuit voltages.

   OBSERVATION V_A = _____ V. V_{R_3} = _____ V.
                                                (across the open)

   V_{R_1} = _____ V. V_{R_4} = _____ V.

   V_{R_2} = _____ V.

   CONCLUSION   Opening a parallel branch in a series-parallel circuit causes I_T to (*increase, decrease*) _____ but not to _____ mA. The voltage drops across the components in series with the main line (*increase, decrease*) _____, whereas the voltage across the components directly in parallel with the open (*increase, decrease*) _____. No matter whether an open occurs in the series or parallel sections of the series-parallel circuit, the circuit total resistance will (*increase, decrease*) _____.

Optional Step

6. Use the circuit of Figure 25-1. With the power off, remove R_4 from the circuit to simulate R_4 opening. Turn the power back on and measure and record V_A and the voltage across the open terminals where R_4 was previously connected.

   ⚠ OBSERVATION V_A measures _____ V. V_{R_4} measures _____ V.

   ⚠ CONCLUSION   Did the voltage across the opened resistor exactly equal V_A? _____. If not, was it higher or lower than V_A? _____. Explain what might cause this to be true. _____

   _____.

Series-Parallel Circuits
Effects of a Short in Series-Parallel Circuits

PROJECT 26

Name: James Quinn   Date: _____

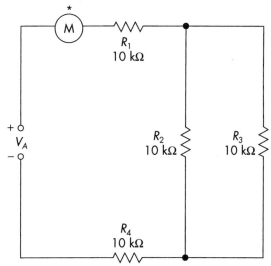

* Use best range to measure currents between 0–1 mA

FIGURE 26-1

PROJECT PURPOSE To demonstrate, through measurements and calculations, the various effects on electrical parameters of a short occurring in a series portion and in a parallel portion of a series-parallel circuit. To note the effect on the overall circuit total current when a short occurs.

PARTS NEEDED ☐ DMM ☐ VVPS (dc) ☐ CIS ☐ Resistors 10 kΩ (4)

PROCEDURE

1. Connect the initial circuit as shown in Figure 26-1.

2. Increase V applied until 0.5 mA of current is indicated by the milliammeter. Measure and note the circuit voltages.

   ⚠ OBSERVATION
   V_A = 12.28 V. V_{R_3} = 2.4 V.
   V_{R_1} = 4.91 V. V_{R_4} = 4.9 V.
   V_{R_2} = 2.4 V.

PART 5: Series-Parallel Circuits

3. Calculate the current through each component using Ohm's law.

   OBSERVATION I_{R_1} = __.49__ mA. I_{R_3} = __.24__ mA.

   I_{R_2} = __.24__ mA. I_{R_4} = __.49__ mA.

   CONCLUSION Total current flows through R_1 and R __4__. One- (*third, half, fourth*) __half__ of total current flows through R_2 and one-half through R (*1, 2, 3, 4*) __3__.

4. Simulate R_1 shorting by replacing it with a jumper wire. Measure the circuit voltages and record.

   OBSERVATION V_A = __12.3__ V. V_{R_3} = __4.1__ V.

   V_{R_1} = __0__ V. V_{R_4} = __8.18__ V.
   (across the short)

   V_{R_2} = __4.1__ V. I_T ≈ .8

   CONCLUSION Shorting a component in series with the source caused I_T to (*increase, decrease*) __Increase__, the voltage drops across all the "unshorted" or normal components to (*increase, decrease*) __Increase__, and the voltage across the shorted element to (*increase, decrease*) __decrease__ to __0__ V. Do Kirchhoff's voltage and current laws still hold true when there is a short in the circuit? __yes__.

5. Replace R_1 into the circuit. Now replace R_3 with a short. Measure and record the circuit voltages.

   OBSERVATION V_A = __12.3__ V. V_{R_3} = __0__ V.

   V_{R_1} = __6.15__ V. V_{R_4} = __6.13__ V.

   V_{R_2} = __0__ V. I_T ≈ .6

   CONCLUSION Shorting a parallel branch caused I_T to (*increase, decrease*) __decrease__ to __.6__ mA. The voltage across the shorted branch and any branch in parallel with it (*increases, decreases*) _____ to __0__ V. The voltages across all other components (*increased, decreased*) _____ because total current (*increased, decreased*) _____. A short anywhere in any type of circuit will cause the circuit total resistance to (*increase, decrease*) _____. Therefore, the circuit total current will (*increase, decrease*) _____.

Story Behind the Numbers
Series-Parallel Circuits

Name: _____   Date: _____

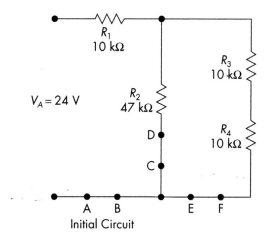

Initial Circuit

Procedure

NOTE When performing the procedure steps, you have the options of using a calculator or using an Excel spreadsheet program "worksheet" for any required calculations. You may also use Excel for creating tables and for generating graphs.

1. In this project you will connect and use the *Initial Circuit,* shown above, plus two modifications of the *Initial Circuit (Modified Circuit 1* and *Modified Circuit 2).* You will make measurements and observations, and create a Data Table from which you will develop graphs and perform analyses.

2. Connect the circuit as shown **without** the power source connected to the circuit.

   NOTE ▶ Identifier dots A-B, C-D, and E-F are simply points that are identified for inserting current meters in later steps.

3. Without the power source connected to the circuit, measure the circuit total resistance (R_T) for this *Initial Circuit.* Record this total resistance value in the Data Table, where indicated.

4. Temporarily place a jumper wire across R_1. This is the configuration for *Modified Circuit 1.* (It simulates the initial circuit without R_1 being part of the circuit.) Measure the R_T of this *Modified Circuit 1.* Record the circuit total resistance value in the appropriate location on the Data Table.

5. After obtaining the total resistance measurement for *Modified Circuit 1,* remove the jumper wire from across R_1. Now, place a jumper wire across R_3, which simulates R_3 being removed from the *Initial Circuit* and replaced by a wire. This configuration of the circuit represents *Modified Circuit 2.*

Measure and record the R_T of this circuit configuration in the appropriate location for *Modified Circuit 2* in the Data Table. After recording this total resistance value, remove the jumper from across R_3 to return the circuit to the *Initial Circuit* configuration.

6. Before connecting the power source to the *Initial Circuit* configuration, break the circuit between points A and B and insert a current meter, observing proper polarity of connections. If two more meters are available, it would be good to break the circuit wiring between points C and D, and E and F, and insert a current meter at each of these locations, as well. This will allow you to monitor current levels at all three locations simultaneously without having to turn the power off and insert meters one at a time at each location in order to measure the various current levels.

7. Connect the power supply to the circuit; then, adjust for 24 volts of applied voltage to the *Initial Circuit* configuration. Make all measurements and calculations required to finish filling in the Data Table for the *Initial Circuit* data, as indicated.

8. Modify the circuit to the *Modified Circuit 1* configuration by placing a jumper lead across R_1. Now, make all measurements and calculations required to completely fill in the data asked for in the *Modified Circuit 1* column of the Data Table.

9. Remove the jumper from across R_1 and place the jumper across R_3. This is now the configuration called for to make *Modified Circuit 2*. Make all measurements and calculations required to completely fill in the Data Table for *Modified Circuit 2*, as appropriate.

10. After completing all the data required in the Data Table, use this data to create three separate bar graphs that illustrate the voltage drops across each resistor for each circuit configuration, respectively. The bar charts' **x-axes** should identify the four resistors (R_1, R_2, R_3, and R_4) and their resistance values. The **y-axis** of each graph will represent the voltage values.

    HINT ➤ If you set up the table in Excel, the automated chart creation functions will greatly simplify creation of the requested charts.

11. Use the MultiSIM program and simulate the "Loaded Voltage Divider Circuit" called for in textbook Chapter 6: Practice Problems 6. Be sure to change the circuit "Labels" and "Values" for the components in the MultiSIM simulation circuit to match the labels and values you calculated in answering the problem. Once the circuit is appropriately set up in MultiSIM, check the currents through R_{L_1}, R_{L_2}, and R_{L_3}. Also, check the voltages at test points *A*, *B*, and *C* with respect to ground reference. Record all your measurements in the Data Table, as indicated.

    NOTE ➤ Did the MultiSIM simulated circuit measurements confirm your calculated values for Practice Problems 6?

12. After completing the Data Table and producing the required bar graphs, answer the Analysis Questions and create the brief Technical Lab Report to complete the project.

Data Table

V_A to be 24 V for all circuits used in this project.

Identifiers	Initial Circuit Color-Coded & Calculated Values	Initial Circuit Color-Coded & Measured Value (approx.)	Modified Circuit 1 (R_1 Replaced by a Jumper) Color-Coded & Measured Values (approx.)	Modified Circuit 2 (R_3 Replaced by a Jumper) Color-Coded & Measured Values (approx.)
*R_T in kΩ				
*R_1 in kΩ			NA	
*R_2 in kΩ				
*R_3 in kΩ				NA
*R_4 in kΩ				
	Calculated Values	**Measured Values**	**Measured Values**	**Measured Values**
I_T in mA				
I_{R_1} in mA			NA	
I_{R_2} in mA				
I_{R_3} in mA				NA
I_{R_4} in mA				
V_A in V				
V_{R_1} in V			NA	
V_{R_2} in V				
V_{R_3} in V				NA
V_{R_4} in V				
P_T in mW				
P_{R_1} in mW			NA	
P_{R_2} in mW				
P_{R_3} in mW				NA
P_{R_4} in mW				
	MultiSIM Measured Values			
I_{RL_1} (mA)		—	—	—
I_{RL_2} (mA)		—	—	—
I_{RL_3} (mA)		—	—	—
Point A (V)		—	—	—
Point B (V)		—	—	—
Point C (V)		—	—	—

***NOTE:** *DO NOT* have power supply connected to circuit when making R measurements!

Analysis Questions

NOTE Answers to these Analysis Questions should be clearly numbered and documented on separate sheets of paper with your name and the date at the top of each page. These answer sheets are to be turned in with the rest of the project documentation, as appropriate.

1. Which resistor in the *Initial Circuit* is in series with the source?

2. Specifically describe the location of the inserted current meter that reads the same value of current as that which passes through R_1 in the *Initial Circuit*.

3. Which circuit resistor current is measured by the current meter inserted between points **C** and **D**?

4. The current meter inserted between points **E** and **F** is reading the current through which resistor(s)?

5. Explain why the voltage drop across R_1 is less than that across R_2, even though total current passes through R_1 and only a portion of total current passes through R_2.

6. Explain why the voltage drops across R_3 and R_4 are the same in the *Initial Circuit*. Would this be true for any value of applied voltage? Explain.

7. Since R_3 and R_4 are each 10-kΩ resistors and R_1 is also a 10-kΩ resistor, why are their electrical parameters different?

8. List **all** the factors you think are involved in determining a specific component's electrical parameters in a series-parallel circuit.

9. Observing your data and/or graphs, which modification of the *Initial Circuit* had the greatest impact on the electrical parameters of the remaining resistors? Explain why this is true.

10. From your collected data, perform the required calculations to determine what percentage of applied voltage was felt by R_2 in the *Initial Circuit* configuration. In the *Modified Circuit 1* configuration? In the *Modified Circuit 2* configuration?

11. Analyze the data in the Data Table and note any special relationship you see between the three circuits' total resistance values and their circuit total power values. Explain why a change in total circuit resistance affects the circuit total power as it does. Be specific.

Technical Lab Report

Write a brief technical lab report summarizing the technical facts learned from this project. The report should be organized to provide the following:

1. An introductory paragraph describing the type of circuit being analyzed and the key parameters that will be discussed relating to this circuit.

2. A section describing the most important characteristics of this type of circuit that were shown via the collected data in the table and graphs.

3. Any special facts or characteristics about this type of circuit that were highlighted in answering the Analysis Questions.

4. A practical example of how the information learned in this project might help you in operating, troubleshooting, error analysis, or adjusting a circuit of this type in your home setting, in your training program setting, or in a job setting in the real world.

5 A summary statement listing the most positive aspects of the project and any parts of the project that were difficult because of equipment problems or unclear instructions. Include areas that might be improved.

Summary
Series-Parallel Circuits

Name: _____ Date: _____

Complete the following review questions, indicating the appropriate response by placing a check in the box next to the correct answer.

1. The total resistance (R_T) of a series-parallel circuit is always greater than the sum of the Rs through which total current flows.
   - ☐ True
   - ☐ False

2. Total current passes through every component in a series-parallel circuit.
   - ☐ True
   - ☐ False

3. Considering only the components in series with the main line, the resistor that will dissipate the most power is
   - ☐ the largest R
   - ☐ the smallest R
   - ☐ neither of these

4. Considering only the components in the parallel sections of the series-parallel circuit, the component that will dissipate the most power is
   - ☐ the largest R
   - ☐ the smallest R
   - ☐ neither of these

5. In a series-parallel circuit, does the largest resistor always have the largest $I \times R$ drop?
   - ☐ Yes
   - ☐ No

6. In a series-parallel circuit, if a component in series with the source (in the main line) opens, the voltage across the other components will
   - ☐ increase
   - ☐ decrease
   - ☐ remain the same

7. If a resistor in a parallel section of a series-parallel circuit opens, the voltage across the unopened components will
   - ☐ increase in some cases and decrease in others
   - ☐ decrease in all cases
   - ☐ increase in all cases

8. If a component in series with the main line shorts, the voltage across all the other components will
   - ☐ increase
   - ☐ decrease
   - ☐ remain the same

9. If a resistor in a parallel section of a series-parallel circuit shorts, the voltage across all the other components will
   ☐ increase in some cases and decrease in others
   ☐ decrease in all cases
   ☐ increase in all cases

10. To analyze a series-parallel circuit, it is necessary to use both series circuit rules and parallel circuit rules.
    ☐ True
    ☐ False

Student Log
For Optional Troubleshooting Exercise

Name: _____ Date: _____

NOTE ➤ Complete instructions are located on page xiii of the Preface. It is important that you read these instructions carefully before performing any troubleshooting exercises.

Students should write down the starting point symptom given to them by the instructor, and then proceed to log each step used in the troubleshooting process. For example, the first step after learning the starting point symptom information is to identify the area to be bracketed as the "area of uncertainty," and should include all circuitry components that "might" cause the trouble producing the symptoms. The next step is to make a decision about the first test, and so on until the exercise is complete and the suspected trouble is determined.

- Starting point symptom

  _____

- Components and circuitry included in "initial brackets" ("area of uncertainty")

  _____

- First test description (what type of test and where?)

  _____

- Components and circuitry *still* included in "bracketed" area after first test

  _____

- Second test description (what type of test and where?)

  _____

- Components and circuitry *still* included in "bracketed" area after second test

  _____

- Third test description (what type of test and where?)

  _____

- Components and circuitry *still* included in "bracketed" area after third test

  _____

- Fourth test description (what type of test and where?)

  _____

- Components and circuitry *still* included in "bracketed" area after fourth test

___

- Fifth test description (what type of test and where?)

___

NOTE ▶ Keep repeating the testing, bracketing, and logging procedure until you are quite sure you have found the trouble. Call the instructor to check your work; then, replace the identified faulty component to see if the circuit operates properly.

- Suspected trouble

___

- Trouble verified by instructor and by circuit operation: ____ Yes ____ No

BASIC NETWORK THEOREMS

PART 6

Objectives

You will connect circuits illustrating Thevenin's, Norton's, and Maximum Power Transfer Theorems.

In completing these projects, you will connect circuits, make measurements, perform calculations, draw conclusions, and be able to answer questions about the following items related to these theorems:

- R_{TH}
- V_{TH}
- R_N
- I_N
- Current through a load
- Voltage across a load
- Efficiency of circuit at maximum power transfer

Project/Topic Correlation Information

PROJECT	TEXT CHAPTER	SECTION	RELATED TEXT TOPIC(S)
27 Thevenin's Theorem	7	7-4	Thevenin's Theorem
28 Norton's Theorem	7	7-5 7-6	Norton's Theorem
29 Maximum Power Transfer Theorem	7	7-2	Maximum Power Transfer Theorem

Basic Network Theorems
Thevenin's Theorem

PROJECT 27

Name: _____   Date: _____

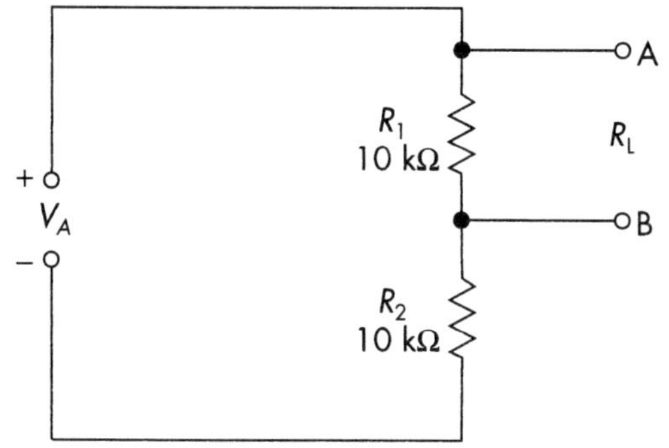

FIGURE 27-1

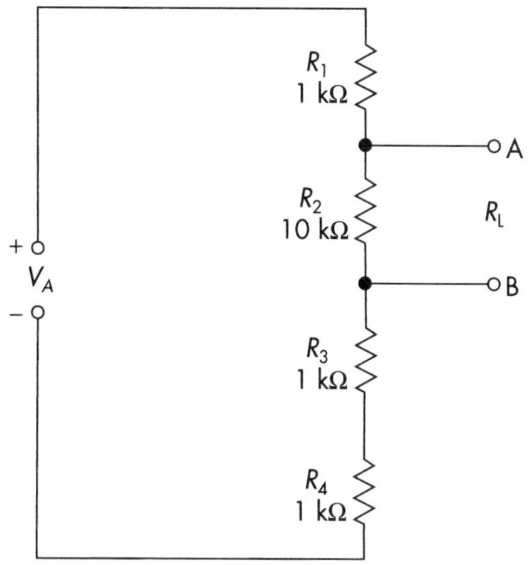

FIGURE 27-2

PROJECT PURPOSE To practice analyzing a resistive network, using Thevenin's theorem. To calculate R_{TH} and V_{TH} and predict circuit parameters based on these values. To confirm these predictions by actual measurements.

PARTS NEEDED
- ☐ DMM
- ☐ VVPS (dc)
- ☐ CIS
- ☐ Resistors
  1 kΩ (3)   10 kΩ (2)
  4.7 kΩ

SPECIAL NOTE:

For our purposes, Thevenin's theorem might be stated as follows: The current through a load resistor (R_L) connected between any two points on an existing resistive network (circuit) can be calculated by dividing the voltage at those two points, prior to connecting R_L, by the sum of R that would be measured at those two points (with the voltage source(s) replaced by shorts) + R_L. Therefore: $I_L = V_{TH}/R_{TH} + R_L$.

In essence, Thevenin's theorem tells us that the entire network of resistors and voltage source(s) can be replaced by an equivalent circuit of a voltage source, called V_{TH}, and a single series resistor, called R_{TH}.

V_{TH} = the open-circuit voltage at the two points (V without R_L connected).
R_{TH} = resistance seen when looking back into the circuit at the two points, with any sources replaced by shorts (and without R_L connected).

In order to quickly calculate I_L (current through R_L) when R_L is to be connected to any existing network (complicated or otherwise), we then consider R_L as being connected across the Thevenin equivalent circuit so the circuit consists of V_{TH} supplying voltage to R_{TH} and R_L in series. Therefore: $I_L = V_{TH} / R_{TH} + R_L$.

PROCEDURE

1. Connect the initial circuit as shown in Figure 27-1.

2. Apply the voltage distribution characteristics of series circuits and determine what the voltage would be between points A and B with a source V of 20 volts.

   OBSERVATION $V_{A-B} =$ _____ V.

   CONCLUSION   In a series circuit consisting of two equal-value resistors, each resistor will drop (1/3, 1/2, 1/4) _____ of V applied. If the load resistor, R_L, is to be connected to points A and B, then $V_{TH} =$ _____ V.

3. Assume the source was zero ohms and calculate R_{TH}.

   NOTE ▶ R_1 and R_2 would then be effectively in parallel.

   OBSERVATION $R_{TH} =$ _____ Ω.

   CONCLUSION   Therefore, the Thevenin equivalent circuit would be drawn as follows:

   $R_{TH} =$ _____
   $V_{TH} =$ _____ V

4. If R_L is to be a 10-kΩ resistor, calculate I_L and V_L.

   OBSERVATION $I_L =$ _____ mA. $V_L =$ _____ V.

   CONCLUSION $V_L = I_L \times$ _____.

5. Actually apply 20 volts of V applied. Measure and record circuit parameters as indicated in the "Observation" section.

   OBSERVATION $V_A =$ _____ V. $V_{R_2} =$ _____ V.
   $V_L =$ _____ V.

   CONCLUSION   With R_L connected, total circuit current is _____ mA. If R_L were disconnected, I_T would = _____ mA and V_2 would = _____ V.

6. Using the Thevenin equivalent circuit, as appropriate, calculate and project what the values of V_L and I_L would be if R_L were changed to a value of 4.7 kΩ.

   OBSERVATION V_L calculated = _____ V. I_L calculated = _____ mA.

7. Change R_L to a value of 4.7 kΩ and make measurements needed to compare measurements with your projections.

   OBSERVATION V_L measured = _____ V.

   CONCLUSION   Did the measured and projected values for V_L reasonably compare? _____. Were the calculations made easier by virtue of the Thevenin equivalent circuit approach? _____.

8. Connect the circuit as shown in Figure 27-2.

9. Calculate V_{TH} and R_{TH} (assume 13 volts V applied).

   OBSERVATION V_{TH} = _____ V. R_{TH} = _____ Ω.

   CONCLUSION   Therefore, the Thevenin equivalent circuit would be drawn as follows:

   R_{TH} = _____

   V_{TH} = _____ V

10. If R_L is to be a 10-kΩ resistor, calculate I_L and V_L.

    OBSERVATION I_L = _____ mA. V_L = _____ V.

    CONCLUSION $I_L = V_{TH}/R_L + R_{TH}$ = _____ mA.

    $V_L = I_L \times R_L$ = _____ × 10 kΩ = _____ V.

11. Apply 13 volts to the circuit. Measure the circuit parameters as indicated in the "Observation" section.

    OBSERVATION V_A = _____ V. V_{R4} = _____ V.

    V_L = _____ V.

    CONCLUSION   The main advantage of using Thevenin's theorem for solving I_L is not that it particularly simplifies the initial calculations for a single value of R_L, but rather, that if R_L were going to be changed several times, V_{TH} and _____ could be used for each case. This means that each new change of R_L does not necessitate solving the whole network problem each time. All that is necessary is to substitute the new value of R_L in the formula: $I_L = V_{TH}/$ _____.

Basic Network Theorems
Norton's Theorem

PROJECT 28

Name: _____ Date: _____

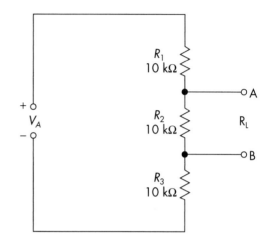

FIGURE 28-1   **FIGURE 28-2**

PROJECT PURPOSE To practice analyzing a resistive network, using Norton's theorem. To calculate R_N and I_N and predict circuit parameters based on these values. To confirm these predictions by actual measurements.

PARTS NEEDED
- ☐ DMM
- ☐ VVPS (dc)
- ☐ CIS
- ☐ Resistors
  - 1 kΩ   10 kΩ (4)
  - 5.6 kΩ

SPECIAL NOTE:

For our purposes, Norton's theorem might be stated as follows: Any two-terminal network of resistors and source(s) may be represented by a single current source shunted by a single resistance. The size of the current source is determined by calculating the amount of current that will flow through the two terminals if they are shorted. The size of the resistance (shunt R) is equal to the resistance that would be seen looking back into the network from the two points (when they are not shorted and without an R_L).

In essence, the above statement means:

- I_N (equivalent Norton current source) = current that would flow through a short at points A and B if A and B were shorted.
- R_N = resistance seen when looking back into the circuit at the two points (to which R_L will be connected), with any sources replaced by shorts. (In effect, R_N is the same as R_{TH} would be.)

In order to quickly calculate I_L (current through R_L) when R_L is to be connected to points A and B, we then consider R_L as being connected across the Norton equivalent circuit so the circuit consists of I_N supplying current to R_N and R_L in parallel and the current dividing according to the rules of parallel circuits. Therefore: $I_L = (R_N/R_N + R_L) \times I_N$.

PROCEDURE

1. Connect the initial circuit as shown in Figure 28-1.

2. Assume a short across points A and B and determine what the current through this short would be if the source voltage is to be 15 volts.

   OBSERVATION $I_N = $ _____ mA.

   CONCLUSION   In effect, R _____ is shorted out and the circuit total resistance is _____ Ω.

3. Assume the source is zero ohms and calculate R_N.

   NOTE ➤ R_1 and R_2 would then be effectively in parallel.

   OBSERVATION $R_N = $ _____ Ω.

   CONCLUSION   The Norton equivalent circuit would therefore be drawn as follows:

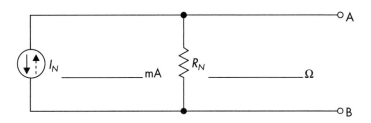

4. If R_L is to be a 10-kΩ resistor, calculate I_L and V_L. (Use the formula for I_L shown in the Special Note.)

   OBSERVATION $I_L = $ _____ mA. $V_L = $ _____ V.

   CONCLUSION $V_L = I_L \times $ _____ .

5. Actually apply 15 volts to the circuit and measure the circuit parameters to verify predicted results.

   OBSERVATION $V_A = $ _____ V. $V_{R_2} = $ _____ V.

   $V_L = $ _____ V.

   CONCLUSION   Does Norton's theorem actually work in analyzing and predicting I_L? _____ .

6. Using the Norton equivalent circuit, as appropriate, calculate and project what the values of V_L and I_L would be if R_L were changed to a value of 5.6 kΩ.

   OBSERVATION V_L calculated = _____ V. I_L calculated = _____ mA.

7. Change R_L to a value of 5.6 kΩ and make measurements needed to compare measurements with your projections.

   OBSERVATION V_L measured = _____ V.

   CONCLUSION Did the measured and projected values for V_L reasonably compare? _____. Were the calculations made easier by virtue of the Norton equivalent circuit approach? _____.

8. Connect the initial circuit as shown in Figure 28-2.

9. Calculate I_N and R_N (assume 20 volts V_A).

   OBSERVATION I_N = _____ mA. R_N = _____ Ω.

   CONCLUSION $I_N = V_A /$ _____ Ω = _____ mA.

   R_N = 10 kΩ × _____ /10 kΩ + _____ = _____.

10. If R_L is to be 1 kΩ, calculate I_L and V_L for a V_A of 20 volts.

    OBSERVATION I_L = _____ mA. V_L = _____ V.

    CONCLUSION $I_L = \left(\dfrac{R_N}{R_N + \underline{}}\right) \times I_N =$ _____ mA.

11. Apply 20 volts to the circuit. Measure parameters and verify the predictions made by calculations using Norton's theorem.

    OBSERVATION V_A = _____ V. V_{R_3} = _____ V.

    V_L = _____ V.

    CONCLUSION The predictions made using Norton's theorem (*were, were not*) _____ verified by the measurements.

12. If R_L were to be changed to a 10-kΩ resistor, use Norton's theorem and predict the parameters called for in the "Observation" section (assume 20 volts V_A).

    OBSERVATION I_N = _____ mA. V_{R_1} = _____ V.

    R_N = _____ Ω. V_{R_2} = _____ V.

    I_L = _____ mA. V_{R_3} = _____ V.

    V_L = _____ V.

    CONCLUSION Changing the value of R_L that will be used (*does, does not*) _____ change the Norton equivalent circuit. The Norton equivalent circuit for the circuit shown on the setup diagram can therefore be drawn as follows:

Basic Network Theorems
Maximum Power Transfer Theorem

Name: James Quinn   Date: 10-30-14

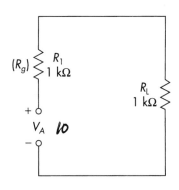

V_A 10

$R_T = 2K$
$I_T = \frac{10}{2K} = 5mA$

FIGURE 29-1

PROJECT PURPOSE To verify the maximum power transfer theorem through circuit measurements and calculations. To observe the relationships of R_g and R_L values to both the amount of power delivered to the load, and the efficiency of power delivery to the load.

PARTS NEEDED
- ☐ DMM
- ☐ VVPS (dc)
- ☐ CIS
- ☐ Resistors
  - 100 Ω   10 kΩ
  - 1 kΩ (2)

SPECIAL NOTE:

For our purposes, the maximum power transfer theorem might be stated as: Maximum power will be transferred from the generator (or source) to the load (R_L) when the resistance of the load equals the resistance of the generator (or source).

$R_T = 2K$ $I = 5mA$
$V = 10$ $P_L = 50mW$

PROCEDURE

1. Connect the initial circuit as shown in Figure 29-1.

2. Apply <u>10</u> volts to the circuit. Measure V_L and calculate P_L.

    ▲ **OBSERVATION** V_L = ~~10~~ 4.9 V. P_L = ~~50mW~~ ~~10~~ 24.9 mW.

    ▲ **CONCLUSION**   This circuit is solved as a simple (<u>series</u>, parallel) __series__ circuit.

3. Assume that R_1 represents the resistance of the generator (R_g). Calculate the power transfer efficiency as follows:

    Efficiency = $(P_{out}/P_{in}) \times 100$ where $P_{out} = P_L$ and $P_{in} = P_T$

PART 6: Basic Network Theorems

$I_T = .9 mA$

⚠ OBSERVATION P_{out} = __75__ mW.   Efficiency = __50__ %.
    P_{in} = __50__ mW.

⚠ CONCLUSION   When $R_L = R_g$, (1/3, 1/4, 1/2) __½__ the total power provided by the source is transferred to the load (R_L).

4. Change R_L to 10 kΩ. Measure V_L and calculate P_L and efficiency (use 10 volts V_A).

    ⚠ OBSERVATION V_L = __9.__ V. P_{in} = __9__ mW.
       P_L = __8.1__ mW.   Efficiency = __90__ %.
       P_{out} = __8.1__ mW.

    ⚠ CONCLUSION   Making R_L larger than R_g caused efficiency to (*increase*, decrease) _____.
    However, P_L was (less, *more*) _____ than when $R_L = R_g$. The reason for this is that, with the higher R_L, the circuit total resistance was larger. Thus, the total power supplied to the circuit by the source was (less, *more*) _____ than when $R_L = R_g$, even though P_L was a (larger, *smaller*) _____ percentage of P_T, the power transferred to the 10-kΩ load was (less, *more*) _____ than to the 1-kΩ load.

5. Change R_L to a 100-Ω resistor. Measure V_L and calculate P_L and efficiency (use 10 volts V_A).

    ⚠ OBSERVATION V_L = __.9__ V. P_{in} = _____ mW.
    $I = 9 mA$ P_L = __8.1__ mW.   Efficiency = _____ %.
       P_{out} = _____ mW.

    ⚠ CONCLUSION   When R_L is smaller than R_g, the efficiency is (more, less) _____ than 50% as when $R_L = R_g$. Total power produced by the source with R_L smaller than R_g is greater than for maximum power transfer conditions. However, a smaller percentage of the power is dissipated by R _____.

Story Behind the Numbers
Basic Network Theorems

Name: _____   Date: _____

[Circuit diagram: (Source) V applied connected to R_1 = 1 kΩ at top, R_4 = 1 kΩ and R_3 = 1 kΩ at bottom, R_2 = 10 kΩ in parallel between points A and B, with R_L connected across A–B]

Procedure

SUGGESTION Review and/or have the key formulas handy for reference as you work through the theorems in this project. For Thevenin's theorem, review formulas and methods for determining I_L, V_{TH}, and R_{TH}. For Norton's theorem, review formulas and methods for determining R_N, I_N, and I_L.

1. Connect the Initial Circuit shown above in order to examine **Thevenin's theorem.**

2. With the power source *disconnected* from the circuit, jumper the points at which the power source will eventually be connected to the circuit. Measure the circuit resistance between points **A** and **B**. What is this resistance value called? What is its value for this circuit?

3. Connect the power source to the circuit and set its output to 13 volts. Measure the voltage between points **A** and **B**. What is this value of voltage called? What is its value in this circuit?

   NOTE ➤ Turn off the power source.

4. Record the values you measured for R and V between points **A** and **B** in the appropriate locations for Thevenin's theorem data in the Data Table.

5. Assume that you will connect an R_L value of 27 kΩ between points **A** and **B**. Using the Thevenin equivalent circuit data, calculate the value of anticipated I_{RL} and V_{RL} you would measure with 13 volts applied to the circuit. Record your calculated value in the appropriate locations in the Data Table.

6. **Turn on the power source** and set its output for 13 volts. Measure the value of V_{RL} and record this measured value in the appropriate location in the Data Table. From the measured V_{RL}, and the color-coded resistor value of R_L, calculate the value of I_{RL} and record it in the Data Table, as appropriate.

   NOTE ➤ Turn off the power source.

7. Modify the *Initial Circuit* as follows in order to examine **Norton's theorem:**
   - Change R_1 to make it a 10-kΩ resistor.
   - Change R_3 to make it a 10-kΩ resistor.
   - Remove R_4 from the circuit and place a jumper between the points to which it was connected.

8. With the power source *disconnected* from the circuit, jumper the points at which the power source will eventually be connected to the circuit. Measure the circuit resistance between points **A** and **B**. For Norton's theorem, what is this resistance value called? What is its value for this circuit?

9. Connect a current meter between points **A** and **B**. Connect the voltage source to the circuit and set its output to 20 volts. Measure the current value. For Norton's theorem, what is the identifier used for this current? Record the measured value of R_N from step 8 and the measured value of current between points **A** and **B** (I_N) in the appropriate locations in the Data Table for the Norton's theorem data.

   NOTE ➤ Turn off the power source.

10. Assume that you will connect an R_L value of 1 kΩ between points **A** and **B**. Using the Norton equivalent circuit data and appropriate Norton formulas, calculate the value of anticipated I_{RL} and V_{RL} you would measure with 20 volts applied to the circuit. Record your calculated value in the appropriate locations in the Data Table.

11. Actually connect a 1-kΩ R between points A and B, then **turn on the power source** and set its output for 20 volts. Measure the value of V_{RL} and record this measured value in the appropriate location in the Data Table. From the measured V_{RL} and the color-coded resistor value of R_L, calculate the value of I_{RL} and record in the Data Table, as appropriate.

    NOTE ➤ Turn off the power source.

12. Modify the *Initial Circuit* as follows to examine the **Maximum Power Transfer Theorem:**
    - Remove resistors R_3 and R_4 from the *Initial Circuit* and replace with jumpers to complete the circuit, as appropriate.
    - Use 10-kΩ resistors for both R_1 and R_2.

13. Let R_1 represent the resistance of the generator (R_g) and R_2 represent the value of R_L. Apply 20 volts to the circuit. Measure V_L and calculate the value of P_L. Record these values in the appropriate locations in the Data Table.

    NOTE ➤ Turn off the power source.

14. Change the value of R_L, (R_2) sequentially to the values called for in the Data Table. With each change in R_L value, apply 20 volts to the circuit. Measure V_L and calculate the value of P_L for each case. Record these values in the appropriate locations in the Data Table.

    NOTE ➤ Turn off the power source.

15. Referring to the Maximum Power Transfer data in the Data Table, create a line graph that shows the relationship of the power transferred to the load resistance (R_L) versus the value of the load resistance.

HINT ▶ Don't forget to use the advantages of Excel for your Data Table calculations and graph creation, if you can.

16. After completing all the data called for in the Data Table, answer the Analysis Questions and produce a brief Technical Lab Report to complete the project.

Data Table

Identifiers	Color-Coded & Calculated Values	Measured Values (Approximate)
For Thevenin's Theorem		
$R_1 = 1$ kΩ		—
$R_2 = 10$ kΩ		—
$R_3 = 1$ kΩ		—
$R_4 = 1$ kΩ		—
$V_A = 13$ V	—	
$R_{thevenin}$ (kΩ)		
$V_{thevenin}$ (V)		
R_L (color code) (kΩ)		
I_{RL} (mA)		
V_{RL} (V)		
Thevenin Equivalent Circuit: $V_{TH} =$ V; $R_{TH} =$ kΩ in series with $R_l =$ kΩ		
For Norton's Theorem		
$R_1 = 10$ kΩ		—
$R_2 = 10$ kΩ		—
$R_3 = 10$ kΩ		—
$R_4 =$ not in circuit	—	—
$V_A = 20$ V	—	
R_{norton} (kΩ)		
I_{norton} (mA)		
R_L (color code) (kΩ)		
I_{RL} (mA)		
V_{RL} (V)		

Data Table (continued)

Identifiers	Color-Coded & Calculated Values	Measured Values (Approximate)
Norton Equivalent Circuit: I_N = mA; R_N = kΩ in parallel with R_L = kΩ		
For Maximum Power Transfer Theorem		
R_g (R_1) = 10 kΩ		—
R_{L1} = 1 kΩ		—
R_{L2} = 4.7 kΩ		—
R_{L3} = 10 kΩ		—
R_{L4} = 27 kΩ		—
R_{L5} = 47 kΩ		—
V_A = 20 V	—	
V_{L1} (V)		
V_{L2} (V)		
V_{L3} (V)		
V_{L4} (V)		
V_{L5} (V)		
P_{L1} (mW)		—
P_{L2} (mW)		—
P_{L3} (mW)		—
P_{L4} (mW)		—
P_{L5} (mW)		—

Analysis Questions

NOTE Answers to these Analysis Questions should be clearly numbered and documented on separate sheets of paper with your name and the date at the top of each page. These answer sheets are to be turned in with the rest of the project documentation, as appropriate.

1. Which, if any, of the Thevenin *equivalent circuit* paramenters would change if there were going to be a change in circuit (network) applied voltage?

2. What item(s) or parameters would have to change in a circuit or network being evaluated that would cause a change in the R_{TH} value of the Thevenin equivalent circuit for that circuit?

3. Describe what you think is the key advantage of using the Thevenin theorem for analyzing a circuit or network of resistive components.

4. Once the Thevenin equivalent circuit has been determined for any given network and applied voltage, is it necessary to recalculate the equivalent circuit parameters if the load resistance value is changed? Explain.

5. When using Thevenin's theorem, describe the specific type of equivalent circuit configuration and type of circuit analysis used to find I_L and V_L for the load.

6. Which, if any, of the Norton *equivalent circuit* parameters would change if there was going to be a change in circuit (network) applied voltage?

7. Are there any items in a circuit that, if changed, would affect the value of R_N in a Norton equivalent circuit? If so, which item or items could cause this change?

8. When using Norton's theorem, describe the specific type of equivalent circuit configuration and type of circuit analysis used to find I_L and V_L for the load.

9. Does changing the value of R_L change the Norton basic equivalent circuit in any way?

10. Describe what you think is the key advantage of using the Norton theorem for analyzing a circuit or network of resistive components.

11. Observing the data you collected and the graph you made when examining the Maximum Power Transfer Theorem, at what value of R_L was maximum power transferred to the load? What relationship does this value have to the value of R_g?

12. Observe the data recorded in the Data Table for the Maximum Power Transfer Theorem and from this data calculate the efficiency factor for each of the load resistances used.

    NOTE ▶ You will need to calculate the total power supplied by the source (P_{in}) for each case and the power delivered to the load (P_{out}) for each case. Then, use the formula: Efficiency = $P_{out}/P_{in} \times 100$.

13. What was the circuit efficiency when the maximum power was delivered to the load?

14. Explain why, when a greater percentage of the source power was delivered to the load (higher circuit efficiency), the load received and dissipated less power than when the efficiency was at the optimum power transfer level.

15. When the load was of such a value as to cause less efficiency than the optimum power transfer level, was the source delivering more or less power than at the optimum power transfer situation? Where was the larger share of power being dissipated?

Technical Lab Report

Write a brief technical lab report summarizing the technical facts learned from this project. The report should be organized to provide the following:

1. An introductory paragraph describing the type of circuit being analyzed and the key parameters that will be discussed relating to this circuit.

2. A section describing the most important characteristics of this type of circuit that were shown via the collected data in the tables and graphs.

3. Any special facts or characteristics about this type of circuit that were highlighted in answering the Analysis Questions.

4. A practical example of how the information learned in this project might help you in operating, troubleshooting, error analysis, or adjusting a circuit of this type in your home setting, in your training program setting, or in a job setting in the real world.

5. A summary statement listing the most positive aspects of the project and any parts of the project that were difficult because of equipment problems or unclear instructions. Include areas that might be improved.

Summary
Basic Network Theorems

Name: _____ Date: _____

Complete the following review questions, indicating the appropriate response by placing a check in the box next to the correct answer.

1. To find R_{TH}, we look back into the network from the two points to which R_L will be connected and assume the source(s) as
   - ☐ open
   - ☐ shorted
   - ☐ neither of these

2. When solving for R_{TH}, we assume that R_L is
   - ☐ connected to the circuit
   - ☐ not connected to the circuit

3. When solving for R_{TH}, we assume the part of the circuit connected to the two points to which R_L will be connected to be
   - ☐ open
   - ☐ shorted
   - ☐ neither of these

4. A Thevenin equivalent circuit consists of
   - ☐ a voltage source and series R
   - ☐ a current source and shunt R
   - ☐ neither of these

5. To find R_N, we look back into the network from the two points to which R_L will be connected and assume the source(s) as
   - ☐ open
   - ☐ shorted
   - ☐ neither of these

6. When solving for I_N, we assume the part of the circuit connected to the two points to which R_L will be connected to be
   - ☐ open
   - ☐ shorted
   - ☐ neither of these

7. A Norton equivalent circuit consists of
   - ☐ a voltage source and series R
   - ☐ a current source and shunt R
   - ☐ neither of these

8. Maximum power transfer from source to load occurs when
   - ☐ R_L is greater than R_g
   - ☐ R_L is less than R_g
   - ☐ $R_L = R_g$
   - ☐ all are the same

9. Highest efficiency occurs when
   - ☐ R_L is greater than R_g
   - ☐ R_L is less than R_g
   - ☐ $R_L = R_g$
   - ☐ makes no difference

10. At maximum power transfer, the efficiency is
    - ☐ 100%
    - ☐ 50%
    - ☐ 25%
    - ☐ none of these
    - ☐ 10%

NETWORK ANALYSIS TECHNIQUES

PART 7

Objectives

In this series of three projects, you will connect several resistive circuits in order to make measurements and observations regarding network analysis techniques.

In completing these projects, you will connect circuits, make measurements, perform calculations, draw conclusions, and be able to answer questions about the following items relating to network analysis techniques:

- Appropriate labeling of schematic diagrams for mesh and nodal analysis
- Use of equations in solving for circuit parameters using mesh and nodal analysis
- Use of equations when performing wye and delta network conversions

Project/Topic Correlation Information

PROJECT	TEXT CHAPTER	SECTION	RELATED TEXT TOPIC(S)
30 Loop/Mesh Analysis	8	8-1	Assumed Loop or Mesh Current Analysis Approach
31 Nodal Analysis	8	8-2	Nodal Analysis Approach
32 Wye-Delta and Delta-Wye Conversions	8	8-3	Conversions of Delta and Wye Networks

Network Analysis Techniques
Loop/Mesh Analysis

PROJECT 30

Name: _____ Date: _____

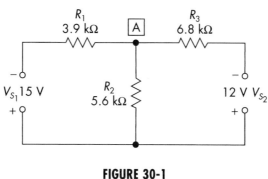

FIGURE 30-1

PROJECT PURPOSE To provide practice in using the loop/mesh analysis technique, and to confirm the accuracy of the technique through actual circuit measurements.

PARTS NEEDED
- ☐ DMM
- ☐ VVPS (dc) (2)
- ☐ CIS
- ☐ Resistors
  - 3.9 kΩ   6.8 kΩ
  - 5.6 kΩ

SPECIAL NOTE:

Steps to use the loop/mesh analysis approach are summarized as follows:

1. Label the assumed mesh currents on the diagram. (Use the arbitrary clockwise direction convention.)
2. Assign mathematical polarities to sources. (That is, the algebraic sign to be used in equations.) If assumed electron-flow mesh current enters positive terminal of source, consider source a positive polarity. If electron-flow mesh current enters negative terminal of source, consider source a negative polarity. For conventional flow, if the assumed mesh current is returning to the source into its negative terminal, the source will be considered positive. If the assumed mesh current is returning into the positive terminal it will be considered negative.
3. Write Kirchhoff's equations for each loop's voltage drops. Assume voltages across resistors in a given loop positive when the voltage drop is caused by its "own mesh" current and negative when caused by an adjacent mesh's assumed current.
4. Solve the resulting equations, using simultaneous equation methods.
5. Verify results using Ohm's law and Kirchhoff's law.

145

PART 7: Network Analysis Techniques

PROCEDURE

1. Draw appropriate arrows and label the meshes on the diagram of Figure 30-1.

   OBSERVATION Number of meshes labeled were _____.

   CONCLUSION The convention used to draw the arrows for assumed mesh currents was to use a (CW, CCW) _____ direction.

2. Assign appropriate polarities to use for sources when performing loop/mesh calculations.

   OBSERVATION $V_{S_1} = (+, -)$ _____. $V_{S_2} = (+, -)$ _____.

   CONCLUSION The reason V_{S_1} was assigned $(+, -)$ _____ polarity is because mesh current enters its $(+, -)$ _____ terminal. V_{S_2} was assigned a $(+, -)$ _____ polarity because mesh current enters its $(+, -)$ _____ terminal.

3. Write Kirchhoff's equations for each loop's voltage drops.

   OBSERVATION Loop A: _____.

   Loop B: _____.

4. Solve equations from step 3 by using simultaneous equation techniques.

   OBSERVATION Equation 1: _____.

   Equation 2: _____.

   $I_A =$ _____. $I_B =$ _____.

5. Use mesh currents and solve for the voltage drops across each resistor.

   OBSERVATION $V_{R_1} =$ _____ V. $V_{R_3} =$ _____ V.

   $V_{R_2} =$ _____ V.

   CONCLUSION Does the sum of $V_{R_1} + V_{R_2}$ equal source V_{S_1}'s value? _____. Does the sum of $V_{R_3} + V_{R_2}$ equal source V_{S_2}'s value? _____.

6. Connect the circuit of Figure 30-1. (Be sure to observe source polarities.)

7. Adjust VVPS 1 so that $V_{S_1} = 15$ V.

8. Adjust VVPS 2 so that $V_{S_2} = 12$ V.

9. Measure voltage drops across each resistor and indicate polarity of each of their voltage drops with respect to point A on the diagram.

   OBSERVATION $V_{R_1} = (+, -)$ _____ V. $V_{R_3} = (+, -)$ _____ V.

   $V_{R_2} = (+, -)$ _____ V.

   CONCLUSION Which resistor's voltage drop represents the result of both meshes' currents? _____.

10. Compare the voltages measured in step 9 with your earlier calculations in step 5.

    OBSERVATION Step 5 V_{R_1} = _____ V.   Step 5 V_{R_3} = _____ V.

    Measured V_{R_1} = _____ V.   Measured V_{R_3} = _____ V.

    Step 5 V_{R_2} = _____ V.

    Measured V_{R_2} = _____ V.

    CONCLUSION Actual circuit operating parameters (*do, do not*) _____ verify the theoretical approach.

11. Calculate the current through each resistor and indicate the direction of current through each R by drawing an appropriate arrow (or dotted-line arrow for conventional current flow) on the diagram next to each R.

    OBSERVATION I_{R_1} = _____ mA. I_{R_3} = _____ mA.

    I_{R_2} = _____ mA.

    CONCLUSION (*Ohm's, Kirchhoff's*) _____ law was used to find current values. Are the calculated currents the same in every case as the "assumed mesh" currents? _____. Why or why not? _____
    _____.

Network Analysis Techniques
Nodal Analysis

PROJECT 31

Name: _____ Date: _____

![Circuit diagram showing two voltage sources and three resistors with node A and reference node]

FIGURE 31-1

PROJECT PURPOSE To provide practice in using the nodal analysis technique and to confirm the accuracy of the technique through actual circuit measurements.

PARTS NEEDED
- ☐ DMM
- ☐ VVPS (dc) (2)
- ☐ CIS
- ☐ Resistors
  - 1.0 kΩ   10 kΩ
  - 5.0 kΩ

SPECIAL NOTE:

The nodal analysis approach steps are summarized as follows:

1. Find the major nodes and select the reference node.
2. Designate currents into and out of major nodes.
3. Write Kirchhoff's Current Law equations for currents into and out of major nodes. (Use V/R statements to represent each current.)
4. Develop equations to solve for voltage across resistor(s) associated with the reference node, and solve for V and I for reference node component(s).
5. Using data accumulated, solve for voltages and currents through other components in the circuit.
6. Verify results using Kirchhoff's Voltage Law.

PART 7: Network Analysis Techniques

PROCEDURE

1. Use arrows (or dotted-line arrows for conventional current flow) and designate currents into and out of the major node on Figure 31-1.

   OBSERVATION *On the diagram:*

   I_1 is current through R _____. I_3 is current through R _____.

   I_2 is current through R _____.

2. Write Kirchhoff's current equation for the loop with the 15-V source.

   OBSERVATION $I_1 + I$ _____ = I _____.

   CONCLUSION   Converting these currents into V/R statements, we can say that V_{R_1}/R_1 + V_R _____ /R _____ equals V_R _____ /R.

   The voltage and current of special interest, due to being associated with the reference node, is V_R ___ and current I ___.

3. Write equations for V across R_2 using loop voltage and current statements.

   OBSERVATION

   For 15-V loop: For 12-V loop:

   $V_{R_1} + V_{R_2} = 15$ V $V_{R_3} + V_{R_2} = 12$ V

   $V_{R_1} = 15 -$ _____. $V_{R_3} = 12 -$ _____.

   CONCLUSION   Voltage across R_2 is result of current I _____.

   Translated into V/R current statements:

   $15\text{ V} - V_{R_2}/10\text{ k}\Omega$ plus $12\text{ V} - V_{R_2}/1\text{ k}\Omega$ equals $V_{R_2}/5\text{ k}\Omega$.

4. Perform appropriate math to solve for V_{R_2}.

   OBSERVATION $V_{R_2} =$ _____ V.

5. Use Ohm's law and solve for I_{R_2}.

   OBSERVATION $I_{R_2} =$ _____ mA.

6. Solve for V_{R_3}, I_{R_3}, V_{R_1}, and I_{R_1}.

   OBSERVATION $V_{R_3} = 12 -$ _____ = _____ V. $V_{R_1} = 15 -$ _____ = _____ V.

   $I_{R_3} =$ _____ V ÷ _____ kΩ. $I_{R_1} =$ _____ V ÷ _____ kΩ.

   equals _____ mA.   equals _____ mA.

7. Verify results of calculations.

   OBSERVATION $V_{R_1} + V_{R_2} = V_{S_1}$ $V_{R_2} + V_{R_3} = V_{S_2}$

   _____ + _____ = 15 V.   _____ + _____ = 12 V.

   CONCLUSION   Did the calculated values using the nodal approach appropriately check out with Kirchhoff's law(s)? _____.

PROJECT 31: Nodal Analysis   151

8. Connect the circuit shown in Figure 31-1. (Be sure to observe source polarities.)

9. Adjust VVPS 1 so that V_{S_1} equals 15 volts. Adjust VVPS 2 so that V_{S_2} equals 12 volts.

10. Measure voltage drops across the resistors.

    ⚠ OBSERVATION V_{R_1} measures _____ V. V_{R_3} measures _____ V.

    V_{R_2} measures _____ V.

    ⚠ CONCLUSION   Do the measured values of V_s reasonably agree with your earlier calculations? _____.

11. Using the measured values of V and the color-coded values of resistors, calculate current through each resistor.

    ⚠ OBSERVATION I_{R_1} = _____ mA. I_{R_3} = _____ mA.

    I_{R_2} = _____ mA.

    ⚠ CONCLUSION   Do the current values calculated on the basis of actual circuit measurements reasonably agree with your previous theoretical calculations? _____.

Network Analysis Techniques
Wye-Delta and Delta-Wye Conversions

PROJECT 32

Name: _____ Date: _____

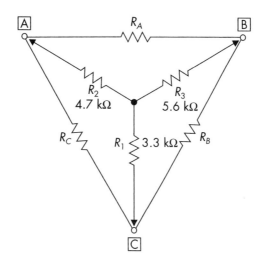

FIGURE 32-1

PROJECT PURPOSE To initially connect a wye network and measure its resistive parameters. Next, to determine the equivalent delta network resistance values through the conversion technique. Finally, to confirm the conversion technique by actual circuit measurements.

PARTS NEEDED
- ☐ DMM
- ☐ CIS
- ☐ Resistors
  - 270 Ω   5.6 kΩ
  - 780 Ω   10 kΩ
  - 820 Ω   12 kΩ
  - 3.3 kΩ   18 kΩ
  - 4.7 kΩ

 SAFETY HINTS Do not connect circuits in this project to any power sources!

SPECIAL NOTE:

The process of converting wye-to-delta and delta-to-wye networks enables selection of appropriate circuit component values that will provide the same electrical load to sources when converting from one of these circuit configurations to the other. For our purposes, we will use the nomenclatures R_1, R_2, and R_3 to designate wye configuration resistive components and R_A, R_B, and R_C to designate delta configuration resistive components. See Figure 32-1.

153

To find values of delta components when the wye component values are known, use:

$$R_A = \frac{(R_1R_2)+(R_2R_3)+(R_3R_1)}{R_1}$$

$$R_B = \frac{(R_1R_2)+(R_2R_3)+(R_3R_1)}{R_2}$$

$$R_C = \frac{(R_1R_2)+(R_2R_3)+(R_3R_1)}{R_3}$$

To find wye component values when the delta component values are known, use:

$$R_1 = \frac{R_B R_C}{R_A + R_B + R_C}$$

$$R_2 = \frac{R_C R_A}{R_A + R_B + R_C}$$

$$R_3 = \frac{R_A R_B}{R_A + R_B + R_C}$$

PROCEDURE

1. Connect the wye portion of the circuit shown in Figure 32-1, using the component values shown.

   OBSERVATION $R_1 =$ _____ kΩ. $R_3 =$ _____ kΩ.

   $R_2 =$ _____ kΩ.

2. Using an ohmmeter, measure and record the resistance readings from point A to B, from point B to C, and from point A to C.

   OBSERVATION A to B = _____ kΩ. A to C = _____ kΩ.

   B to C = _____ kΩ.

   CONCLUSION When measuring from A to B, you are actually measuring the series resistance of R _____ and R _____. The resistance from point B to point C equals the series resistance of R _____ and R _____. The resistance from point A to point C equals the series resistance of R _____ and R _____.

3. Use the appropriate formulas and solve for the values of resistors that would form a delta network that is equivalent to the wye network used in steps 1 and 2.

   OBSERVATION $R_A =$ _____ kΩ. $R_C =$ _____ kΩ.

   $R_B =$ _____ kΩ.

   CONCLUSION The equivalent delta resistor values are (*greater, smaller*) _____ in value than the wye network resistor values.

4. Connect a delta circuit using the approximate (available) resistance values calculated in step 3.

   NOTE ➤ You may have to use more than one resistor for each delta leg in order to use standard, available resistor values.

   ⚠ CONCLUSION Is it necessary to use more than one standard resistor for each leg in this case? _____.

5. Use the space in the "Observation" section and draw a schematic showing your circuit setup, which shows the Rs used for each leg. Label the Rs making up R_A, R_B, R_C, as appropriate. Also label test points A, B, and C on your diagram.

6. Use an ohmmeter and measure the R values found between points A to B, B to C, and A to C.

   ⚠ OBSERVATION A to B measures _____ kΩ.   A to C measures _____ kΩ.

   B to C measures _____ kΩ.

   ⚠ CONCLUSION Do the measured values for this equivalent delta network come close to the values measured between the same test points in the wye network? (*yes, no*) _____. If not, what could cause some of the differences? _____
   _____.

   When measuring R between points A and B, you are measuring the equivalent total R of R_A in parallel with what components? _____. When measuring R between points B and C, you are measuring the equivalent total R of R_B in parallel with what components? _____. When measuring R between points A and C, you are measuring the equivalent total R of R_C in parallel with what components? _____.

Story Behind the Numbers
Network Analysis Techniques

Name: _____   Date: _____

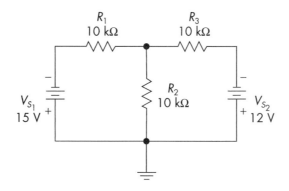

Procedure

1. The circuit diagram above will be used for *both* mesh analysis and nodal analysis portions of this project.

2. First, you should fill in the required data prior to performing a mesh analysis of the above circuit.

   - Label the assumed mesh currents on the diagram. (Use the arbitrary clockwise direction convention.)
   - Assign mathematical polarities to the sources. If assumed electron flow mesh current enters the positive terminal of the source, consider the source a positive polarity. If the electron flow mesh current enters the negative terminal of the source, consider the source a negative polarity. (Note: Use reverse polarities if considering conventional current flow directions.)

3. Write Kirchhoff's equations for each loop's voltage drops.

   Loop A: _____

   Equation 1: _____

   Loop B: _____

   Equation 2: _____

4. Use simultaneous equations and solve the two equations above to find I_A.

157

5. Substitute the value of I_A into Equation 2 (in step 3) and solve for the value of I_B.

6. Use the mesh current values you found and solve for the voltage drops across each resistor in the circuit.

   $I_A =$ _____ $I_B =$ _____

   $V_{R_1} =$ _____ $V_{R_2} =$ _____ $V_{R_3} =$ _____

7. With the circuit connected as shown, measure V_{R_1}, V_{R_2}, and V_{R_3} to see if the measured values correlate well with the theoretically calculated ones from your mesh analysis.

8. Find the major nodes and select the ground reference point as the *reference node*.

9. Use arrows to show currents into and out of the major nodes. Let the current through R_1 be identified as I_1. Let the current through R_2 be identified as I_3. Let the current through R_3 be identified as I_2.

   HINT ➤ Be careful. Renumber identifier I_3 properly while solving the network.

10. On the diagram: I _____ $+ I$ _____ $= I$ _____.

11. Write Kirchhoff's current equation for the loop with the 15-V source.

    NOTE ➤ Convert each current into an appropriate V/R statement: e.g., I_{R_1} would be stated as V_{R_1}/R_1, and so on.

12. Write equations for the voltage across R_2 using loop voltage and current statements.

    For 15-V loop: $V_{R_1} + V_{R_2} = 15$ V; $V_{R_1} = 15 -$ _____.

    For 12-V loop: $V_{R_3} + V_{R_2} = 12$ V; $V_{R_3} = 12 -$ _____.

13. Perform appropriate math to solve for V_{R_2}. Then use Ohm's law to solve for I_2.

14. Solve for:

    V_{R_3} _____, I_{R_3} _____, V_{R_1} _____, and I_{R_1} _____.

15. With the circuit connected as shown, measure: V_{R_1} _____, V_{R_2} _____, and V_{R_3} _____ to see if the measured values correlate well with the theoretically calculated ones from your nodal analysis.

16. After completing all the calculations and measurements data called for thus far, answer the Analysis Questions and produce a brief Technical Lab Report to complete the project.

Analysis Questions

NOTE Answers to these Analysis Questions should be clearly numbered and documented on separate sheets of paper with your name and the date at the top of each page. These answer sheets are to be turned in with the rest of the project documentation, as appropriate.

1. Given the same circuit, voltage sources, and orientations within the circuit, are the parameter analysis results the same for voltage and currents in the circuit using both the mesh and the nodal analyses? In the final analysis, what law is used to calculate the voltages and currents?

2. Describe the difference in characteristics between assumed mesh currents and actual circuit currents.

3. Which type of analysis was used in this project, "generalized loop" definition or "single window frame" definition?

4. Which type of analysis was easier for you in this project, the loop/mesh approach or the nodal approach? Explain why you feel this way.

5. If the values of all three resistors in the circuit were doubled (all other factors remaining the same), would the voltage distribution and voltage values across each resistor remain the same? If your answer to this is yes, then what electrical circuit parameter(s) must have changed and by how much?

Technical Lab Report

Write a brief technical lab report summarizing the technical facts learned from this project. The report should be organized to provide the following:

1. An introductory paragraph describing the type of circuit being analyzed and the key parameters that will be discussed relating to this circuit.

2. A section describing the most important characteristics of this type of circuit that were shown via the collected data in the tables and graphs.

3. Any special facts or characteristics about this type of circuit that were highlighted in answering the Analysis Questions.

4. A practical example of how the information learned in this project might help you in operating, troubleshooting, error analysis, or adjusting a circuit of this type in your home setting, in your training program setting, or in a job setting in the real world.

5. A summary statement listing the most positive aspects of the project and any parts of the project that were difficult because of equipment problems or unclear instructions. Include areas that might be improved.

Summary
Network Analysis Techniques

Name: _____ Date: _____

Complete the following review questions, indicating the appropriate response by placing a check in the box next to the correct answer.

1. When using the loop/mesh analysis technique, assume that mesh currents
   - ☐ should branch at circuit junction points
   - ☐ should not branch at circuit junction points

2. The arbitrary convention regarding directions of assumed mesh currents is that
   - ☐ they should be shown circulating in a clockwise direction through each loop
   - ☐ they should be shown circulating in a counterclockwise direction through each loop

3. Voltage drops across resistors caused by their own assumed mesh currents are
   - ☐ considered negative, when used in mesh equations
   - ☐ considered positive, when used in mesh equations
   - ☐ polarity assignment is not necessary

4. The convention for source polarity assignment(s) indicates that for electron current flow
   - ☐ if the mesh current returns to the source at its positive terminal, the source will be considered negative in equations
   - ☐ if the mesh current returns to the source at its negative terminal, the source will be considered negative in equations
   - ☐ polarity assignments are not necessary for sources

5. A major node, for nodal analysis purposes, is a junction where
   - ☐ two components join
   - ☐ three components join
   - ☐ four components join
   - ☐ none of the above

6. To qualify as a "reference node" in the nodal analysis approach
   - ☐ any node or junction point can be called the reference node
   - ☐ any major node may be designated as the reference node
   - ☐ only the major node with the most junctions can be designated as the reference node

7. The nodal approach is based on
   - ☐ Kirchhoff's Voltage Law
   - ☐ Kirchhoff's Current Law
   - ☐ Ohm's law only

8. Delta networks may also sometimes be called
   - ☐ Tee networks
   - ☐ Pi networks
   - ☐ Vee networks
   - ☐ L networks

9. When converting from a wye network to a delta network,
   - ☐ the numerator in the mathematical equation stays the same for finding all three equivalent delta resistors
   - ☐ the denominator in the mathematical equation stays the same for finding all three equivalent delta resistors
   - ☐ neither the numerator nor the denominator is the same for the three equations used to find the equivalent delta resistors

10. When converting from a delta network to a wye network,
    - ☐ the numerator in the mathematical equation stays the same for finding all three equivalent delta resistors
    - ☐ the denominator in the mathematical equation stays the same for finding all three equivalent delta resistors
    - ☐ neither the numerator nor the denominator is the same for the three equations used to find the equivalent delta resistors

11. When measuring resistance between outside terminal points, which network configuration results in measuring the series resistance of two resistors?
    - ☐ wye network
    - ☐ delta network

12. When measuring resistance between outside terminal points, which network configuration results in measuring the equivalent resistance of parallel resistors?
    - ☐ wye network
    - ☐ delta network

THE OSCILLOSCOPE

PART 8

Objectives

You will adjust a variety of controls on an oscilloscope and connect voltage or signal sources to the scope in order to demonstrate several of the various functions for which an oscilloscope may be used by technicians.

In completing these projects, you will set up and connect an oscilloscope and various voltage/signal source(s), manipulate controls on the equipment, make observations and measurements, draw conclusions, and be able to answer questions about the following items related to the oscilloscope:

- Use of various input/output terminals
- Use of various controls on the front panel
- How to measure voltage(s)
- How to observe waveforms
- How to make phase comparisons
- How to determine frequency

Project/Topic Correlation Information

PROJECT	TEXT CHAPTER	SECTION	RELATED TEXT TOPIC(S)
33 Basic Operation: Familiarization	12	12-1 12-2	Background Information Key Sections of the Scope
34 Basic Operation: Controls Manipulation	12	12-2 12-3	Key Sections of the Scope Combining Horizontal and Vertical Signals to View a Waveform
35 Basic Operation: Vertical Controls and DC V	12	12-2 12-3	Key Sections of the Scope Combining Horizontal and Vertical Signals to View a Waveform
36 Basic Operation: Observing Various Waveforms	12	12-2 12-3	Key Sections of the Scope Combining Horizontal and Vertical Signals to View a Waveform
37 Voltage Measurements	12	12-4	Measuring Voltage and Determining Current with the Scope
38 Phase Comparisons	12	12-5	Using the Scope for Phase Comparisons
39 Determining Frequency	12	12-6	Determining Frequency with the Scope

The Oscilloscope
Basic Operation: Familiarization

PROJECT 33

Name: _____ Date: _____

PROJECT PURPOSE To familiarize you with the most-used controls on an oscilloscope via observation of an actual oscilloscope and through reading its operating manual.

PARTS NEEDED
- ☐ Oscilloscope with cable(s) (a triggered-sweep dual-trace oscilloscope)
- ☐ Equipment manual for oscilloscope used
- ☐ Audio or function generator with cable(s)
- ☐ VVPS (dc)
- ☐ CIS

SPECIAL NOTE:

The oscilloscope is a versatile test instrument that can visually display the relationship between (1) two electrical quantities, and (2) an electrical quantity and time. In this project, you will become acquainted with the basic jacks and controls, and in some cases, screen menus used in obtaining visual waveforms with the scope.

Due to the wide variety of oscilloscopes in use, both in training facilities and in industry, it is not feasible to give precise instructions for every situation. The spectrum of scopes being used may be from the simplest, triggered, dual-trace oscilloscopes to the most sophisticated digital phosphor scopes with automated measurement and math functions appearing on the screen as alphanumeric readouts. Additionally, some scopes may have computer interfacing capability.

Basically, we will try to give generic enough instructions that you can adapt them to the particular scopes you are using at your training facility. It will be important for you, as the student, to acquire the appropriate operator's manual from your instructor for the type of scope you are using to aid you in performing the various tasks in this project and several following projects.

PROCEDURE

1. Obtain the available scope and the appropriate scope operator's manual. Referring to the oscilloscope and appropriate operating manual, locate and point out to your instructor as many of the listed jacks and controls as possible.

    NOTE ▶ Different brands and models may use different names for some of these items. For each item listed, write the name your scope uses that is equivalent to the name we have on our list.

PART 8: The Oscilloscope

⚠ OBSERVATION

Identify the following:

- Power
  - _____(equiv.)

- Intensity
  - _____(equiv.)

- Focus
  - _____(equiv.)

- Ground Jack
  - _____(equiv.)

- AC-GND-DC switch
  - _____(equiv.)

- Vertical Input Jack(s)
  - _____(equiv.)
  - _____(equiv.)

- Vertical Position
  - _____(equiv.)

- Vertical Mode(s)
  - _____(equiv.)
  - _____(equiv.)

- Vertical Gain Control(s)
  - _____(equiv.)
  - _____(equiv.)

- Horizontal Position
  - _____(equiv.)

- Horiz. Time Base Control(s)
  - _____(equiv.)
  - _____(equiv.)

- Trigger Mode(s)
  - _____(equiv.)
  - _____(equiv.)

- Trigger Source(s)
  - _____(equiv.)
  - _____(equiv.)

- Trigger Level
  - _____(equiv.)

⚠ CONCLUSION

Which controls are used to center the display on the scope screen?

_____

_____

_____

Which controls adjust the speed of the horizontal trace?

_____

_____

_____

Which controls help keep the waveform from moving?

_____

_____

_____

Which controls adjust the height of the waveform for a given signal input?

The Oscilloscope
Basic Operation: Controls Manipulation

Name: _____   Date: _____

![Figure 34-1 showing function generator connected to oscilloscope and six waveform displays labeled A through F]

FIGURE 34-1

PROJECT PURPOSE   To provide practice in manipulating oscilloscope controls to achieve desired waveforms.

PROCEDURE

1. Connect the function generator or audio generator to the oscilloscope as shown in Figure 34-1. Adjust the generator frequency to 100 Hz.

   NOTE ➤ Refer to the signal generator operator manual, as required, to learn how to set the output frequencies of the generator.

2. Using the various controls, obtain as many of the waveforms shown on the previous page as possible. Demonstrate waveforms A, B, and D for your instructor.

   NOTE ▶ If you have trouble achieving stable waveforms (that is, "stopping the waveform(s)"), ask the instructor to help you adjust the trigger control(s), as appropriate.

   ⚠ OBSERVATION *Demonstration of waveforms:*

   A _____   D _____

   B _____

   ▶ Instructor initial: _____

   ⚠ CONCLUSION The main controls manipulated to achieve waveforms were the _____ controls and the _____ controls.

   To obtain the waveform shown in D required changing the _____ controls.

3. Disconnect the function generator from the circuit. Make sure the scope is in the dc input mode. Adjust scope controls to obtain a straight horizontal line display that just fills the screen from left to right. (H controls, V and H position controls.)

The Oscilloscope
Basic Operation: Vertical Controls and DC V

PROJECT 35

Name: _____ Date: _____

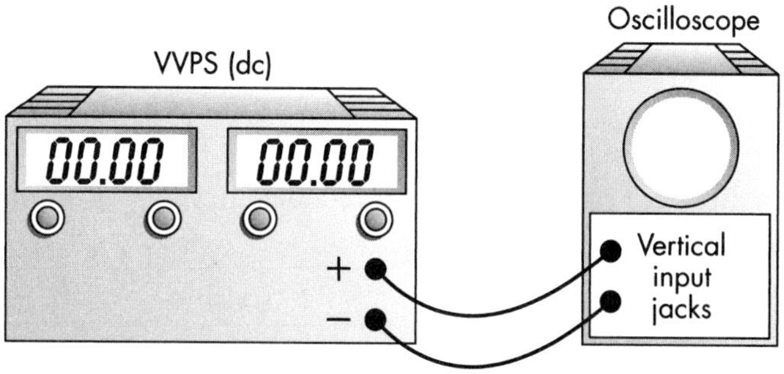

FIGURE 35-1

PROJECT PURPOSE To observe the on-screen results of connecting dc voltage(s) to the vertical input of an oscilloscope. To see the effect of changing, and to practice using, vertical gain control(s).

PROCEDURE

1. Adjust the scope to obtain a straight horizontal line display that just fills the screen from left to right and is centered vertically. Use the horizontal frequency controls and vertical and horizontal position controls, as appropriate.

2. Connect a variable voltage dc source to the vertical input of the scope (– terminal to ground) (+ terminal to vertical input), Figure 35-1.

3. Set the output of the dc source to approximately 5 V.

4. "Make" and "break" the input connection from the dc source to the vertical input jack and observe the display.

   ⚠ OBSERVATION   Does the horizontal trace (line) move when the dc voltage is applied? _____.
   Which direction? _____.

   ⚠ CONCLUSION   The dc voltage applied to the vertical input causes deflection of the scope trace. If the polarity of the input were reversed, would the trace react differently? _____.

   Explain. _____
   _____
   _____.

5. Reverse the polarity of the connections from the dc source to the vertical input of the scope.

   ⚠ OBSERVATION   Does the direction of the display deflection reverse from that shown in step 4? _____.

6. Make appropriate vertical control adjustments to cause the display to "jump" three vertical calibration squares on the scope face "graticule" when the input is connected and disconnected from the scope. When you have made the adjustments, demonstrate this action for your instructor.

   NOTE ➤ When NOT connected, the trace should be in the middle of the screen.

   ⚠ OBSERVATION   Demonstrate that gain controls have been adjusted to achieve specified results.

   ➤ Instructor initial: _____

   ⚠ CONCLUSION   Was more than one control adjusted to achieve the desired results? _____. What were the names of the controls? _____
   _____
   _____.

The Oscilloscope
Basic Operation: Observing Various Waveforms

PROJECT 36

Name: _____ Date: _____

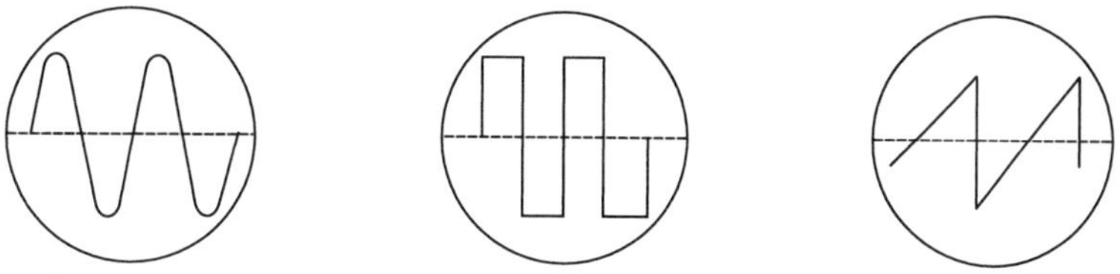

Sine waves   Square waves   Sawtooth or Triangular waves

FIGURE 36-1

PROJECT PURPOSE   To observe various types of waveforms, using the oscilloscope.

PROCEDURE

1. If appropriate signal/function generator(s) are available, demonstrate to your instructor that you can obtain the waveforms shown in Figure 36-1.

   ⚠ **OBSERVATION**   Sine wave; square wave; triangular wave.

   ➤ Instructor initial: _____

173

The Oscilloscope
Voltage Measurements

PROJECT 37

Name: _____   Date: _____

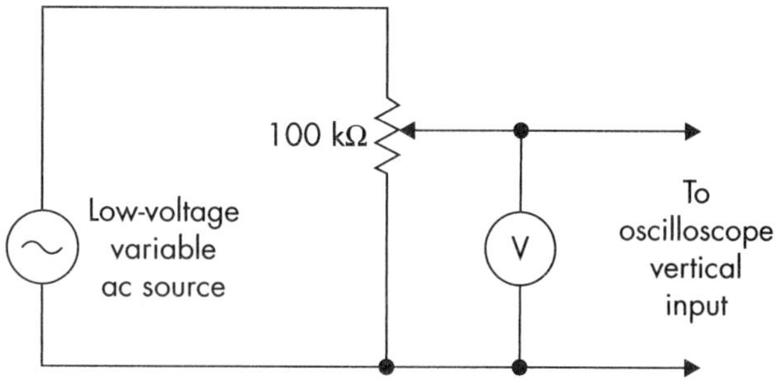

FIGURE 37-1

PROJECT PURPOSE To become familiar with the techniques for measuring voltages with an oscilloscope. To learn to calibrate the scope and interpret the measured values.

PARTS NEEDED
- ☐ Low-voltage variable ac source
- ☐ Oscilloscope
- ☐ DMM
- ☐ CIS
- ☐ 100-kΩ potentiometer
- ☐ Appropriate cables for equipment

SPECIAL NOTE:

The following facts and considerations are important when using the oscilloscope to make voltage measurements:

1. Waveform deflection on the scope indicates the *peak-to-peak* value of the voltage under test.
2. For any **sine wave**, the peak-to-peak deflection on the screen is *directly proportional* to the peak and rms values of the ac applied voltage that produces the peak-to-peak waveform. For example, if the deflection on the CRO screen is 1″ (peak-to-peak deflection) when 1-V rms is applied, it will be 2″ of total deflection when a 2-V rms ac signal is applied. Therefore, it is easy to measure ac voltage values with the scope even though the deflection is peak-to-peak in nature.
3. It is easier to read voltages on the CRO screen when the presentation is a vertical line. To achieve this type of display, set the appropriate horizontal control.

175

PROCEDURE

1. Connect the circuit as shown in Figure 37-1.

2. Monitor input voltage to the scope with the meter and adjust the ac input voltage to 2 volts rms (as indicated by the meter).

3. Adjust the scope positioning and V and H controls to obtain a vertical deflection of **one square** (peak-to-peak).

   > **! CAUTION:** Do not move the vertical V/div or variable controls once this is done, unless directed to do so. However, you can change the vertical position control for ease of viewing.

   ⚠ CONCLUSION One large square of deflection equals _____ V_{rms}; _____ V_P; and _____ V_{P-P}. Each small division on the screen equals _____ V_{rms}.

4. Remove the meter. Adjust the source voltage to obtain two squares of deflection.

   ⚠ OBSERVATION Screen deflection is _____ squares.

   ⚠ CONCLUSION What rms value does this deflection represent? _____ V.

5. Now measure the voltage with the meter.

   ⚠ OBSERVATION Voltage measures _____ V.

   ⚠ CONCLUSION Does this answer agree with the scope measurement? _____.

6. Use the meter and adjust the source voltage to 6 volts rms.

   ⚠ OBSERVATION Amount of deflection equals _____ squares.

   ⚠ CONCLUSION Is the change in deflection essentially linear in nature? _____.

7. If a low-voltage transformer is available, **calibrate** the scope and measure an unknown transformer secondary voltage using the oscilloscope as the measuring device.

   > **! CAUTION:** Do not use exposed 120 Vac connections anywhere in the circuit setup! Be careful! 120 Vac can be lethal!

   ⚠ OBSERVATION Scope is calibrated so that each large square = _____ V_{P-P} or _____ V_{rms}. The unknown voltage is causing _____ large squares of deflection.

   ⚠ CONCLUSION The value of the unknown voltage must equal approximately _____ V_{rms}.

8. Use the meter and verify the measurement taken in step 7 above.

   ⚠ OBSERVATION Meter measures _____ V_{rms}.

   ⚠ CONCLUSION The scope and meter voltage measurements (*are, are not*) _____ close to equal.

The Oscilloscope
Phase Comparisons

PROJECT
38

Name: _____ Date: _____

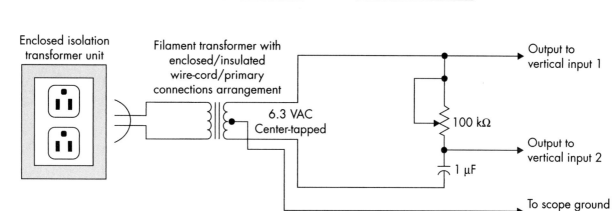

FIGURE 38-1 Variable phase-shift network

FIGURE 38-2 Superimposed waveforms

$$\theta = \frac{360X}{Y}$$

FIGURE 38-3 Calculating phase difference

PROJECT PURPOSE To provide hands-on practice in using the scope with the direct (overlay) phase comparison technique for determining phase difference between signals.

PARTS NEEDED
- ☐ Dual-trace oscilloscope
- ☐ Isolation transformer with outlet sockets
- ☐ *6.3-V filament transformer with insulated connections to power plug
- ☐ 100-kΩ potentiometer
- ☐ 1.0-μF paper/mylar capacitor

*__NOTE:__ Use a low-voltage ac power supply with center-tapped output, if available.

STOP SAFETY HINTS **CAUTION:** Do not use exposed 120 Vac connections anywhere in the circuit setup! Be careful! 120 Vac can be lethal!

SPECIAL NOTE:

In this project, you will briefly look at an approach that may be used to compare the phase of two sine-wave signals (of the same frequency) using an oscilloscope. This method is by direct comparison of signals using a dual-trace scope.

PROCEDURE

1. Connect the variable phase-shift circuit shown in Figure 38-1.

2. Connect the outputs of the network to the dual-trace scope vertical inputs 1 and 2, as appropriate. Adjust the scope H frequency, variable controls, and centering controls to achieve superimposed waveforms, centered on the face of the scope, similar to that shown in Figure 38-2.

   ⚠ OBSERVATION V input 1 gain control set at: _____ V/div.

   V input 2 gain control set at: _____ V/div.

   ⚠ CONCLUSION Are the V levels fed to the two vertical inputs equal? _____.

3. Refer to Figure 38-3. Using the technique shown, determine the phase difference between the two signals you are displaying on your scope.

   ⚠ OBSERVATION Distance X = _____ calibration marks on the scope screen.

   Distance Y = _____ calibration marks on the scope screen.

   ⚠ CONCLUSION The number of degrees difference between the two signals is _____ degrees.

4. Change the setting on the 100-kΩ potentiometer enough to see a noticeable change in the phase and repeat steps 2 and 3 from the previous diagram in Figure 38-1.

   ⚠ OBSERVATION Distance X = _____ calibration marks on the scope screen.

   Distance Y = _____ calibration marks on the scope screen.

   ⚠ CONCLUSION The number of degrees difference between the two signals is _____ degrees.

The Oscilloscope
Determining Frequency

PROJECT 39

Name: _____   Date: _____

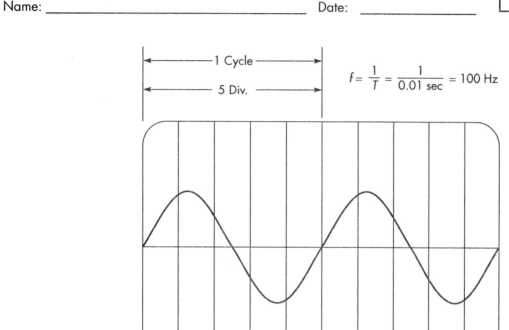

FIGURE 39-1 Scope sweep time setting at 2 ms/div.

PROJECT PURPOSE To provide hands-on practice in determining frequency by using the direct time measurement technique with an oscilloscope.

SPECIAL NOTE:

One method used to determine frequency with an oscilloscope is the direct method. This method uses a scope having a triggered sweep with calibrated sweep times. Since the horizontal sweep is linear and the calibrated sweep provides information about how many milliseconds or microseconds are required for the sweep to travel 1 div. horizontally on the screen, we can easily determine the time for one cycle of the waveform being viewed in Figure 39-1. Once we know the time it takes for one cycle of the waveform (its period), we can then use the $f = 1/T$ formula to find the frequency of the signal causing the pattern. For example, if the calibrated sweep time is set at two milliseconds per division and the scope display shows that one cycle of the signal's waveform starts and finishes in five horizontal divisions, the signal must have a period of 5×2 ms, or 10 ms. The frequency of the signal equals $1/T$, or $1/0.01$ sec = 100 Hz.

PROCEDURE

1. Connect the output of an audio oscillator or function generator to a vertical input on a scope having a calibrated sweep system.

2. Set the sweep for a time of 1 ms per division.

   OBSERVATION   The number of milliseconds for the sweep to travel all the way across the screen is _____ ms.

   CONCLUSION   A signal having a frequency of _____ Hz would cause a waveform display wherein one cycle would take ten divisions on the scope display.

3. Adjust the signal generator frequency to obtain a one-cycle display across ten divisions.

   OBSERVATION   Generator frequency dial calibration reads _____ Hz.

   CONCLUSION   The measured period for this signal is _____ ms. This indicates that the frequency is _____ Hz.

   Does the generator calibration approximately agree with the measured signal frequency? _____ What could cause any differences? _____
   _____
   _____

4. Double the signal input frequency with the scope sweep setting still at 1 ms per division.

   OBSERVATION   How many cycles of the signal are now displayed? _____ cycle(s).

   CONCLUSION   The time for one cycle of this frequency is _____ ms; $f =$ _____ Hz.

   Increasing the frequency of an input signal while keeping the sweep speed the same causes (*more, fewer*) _____ cycles to be displayed.

5. Have the instructor set the frequency of the signal source to a completely different setting. Preferably, the frequency should be set to one that will cause you to change sweep speeds in order to determine the frequency of the signal.

   OBSERVATION   To get a readable display, the sweep speed had to be changed to _____ per division.

   CONCLUSION   The period of one cycle is _____. The frequency determined by the scope's direct measurement system is _____ kHz.

6. Have the instructor check your results in step 5.

   OBSERVATION   *Instructor:* Indicate whether the measurement is correct.

   ____ Yes. ____ No.

7. Practice determining other frequencies, as time permits.

Summary
The Oscilloscope

Name: _____ Date: _____

Complete the multiple-choice questions by placing a check in the box next to the best answer option. Respond to the other types of questions by filling in the blanks or responding appropriately.

1. Four important uses of the oscilloscope are
   - ☐ waveform display, voltage measurement, current measurement, and power measurement
   - ☐ waveform display, voltage measurement, determining frequency, and power measurement
   - ☐ waveform display, voltage measurement, determining phase, and power measurement
   - ☐ waveform display, voltage measurement, determining phase, and determining frequency

2. Why should the intensity or brightness control be set at the lowest point that makes the display readable?
   _____.

3. What oscilloscope terminal is used as the input terminal for the signal whose waveform is to be viewed?
   _____.

4. If an ac voltage of 125 V_{rms} causes 2 div. deflection on the scope screen, what is the peak-to-peak voltage of a signal that causes a deflection of 5 div. with the scope controls unchanged? Show your calculations below.

   Voltage is _____ $V_{P\text{-}P}$.

5. If a frequency of 150 Hz gives a display on the scope screen of three cycles, what is the frequency of a signal that gives a display of four cycles with the scope controls unchanged? Show your calculations below.

   Frequency is _____ Hz.

6. The most common way in which the scope might be used to show phase comparisons between two signals is
   _____.

7. A scope having a calibrated sweep system can be used to directly determine frequency because the time per division of the _____.
   sweep allows us to determine the _____ of the signal being observed.

181

8. What is the frequency of a signal whose display indicates a period of 0.1 ms?
   _____.

9. What is the period of a 25-kHz signal? _____.

10. By which of the following two methods can you more accurately measure a given signal's frequency?
    ☐ Setting the sweep speed so that two cycles of the signal's waveform cover the entire distance of the horizontal sweep
    ☐ Setting the sweep speed so that four cycles of the signal's waveform cover the entire x-axis display

INDUCTANCE

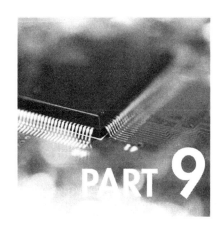

PART 9

Objectives

In these projects you will connect circuits that illustrate the property of an inductor to oppose a change in current and the resultant total inductance of two inductors connected either in series or in parallel.

In completing these projects, you will connect circuits, make measurements, perform calculations, draw conclusions, and be able to answer questions about the following items related to inductance:

- The property of inductance
- Effect of changing value of L
- Total inductance of inductors in series and in parallel

Project/Topic Correlation Information

PROJECT	TEXT CHAPTER	SECTION	RELATED TEXT TOPIC(S)
40 Total Inductance in Series and Parallel	13	13-5	Inductance in Series Inductance in Parallel

SPECIAL NOTES TO STUDENTS AND INSTRUCTORS:

Note 1: Circuit Options

Beginning with *this project* and continuing through the *Series Resonance* and *Parallel Resonance* series of projects there are a number of projects in which two types of circuits are shown. These are a "lower-frequency" circuit and a "higher-frequency" circuit that can be used to perform the project. The instructor has the option of having the students perform each project using either the lower-frequency circuit or the higher-frequency circuit, depending on component availability and instructional program preferences. The instructor may also opt to have students connect and make required measurements and calculations for both types of circuits, as time permits.

Note 2: Measurement Cautions

1. For the *lower-frequency circuit* option, the inductors used in this project and a number of projects that follow are iron-core, filter-choke-type inductors. Under normal operating conditions, this inductor has dc current as well as ac signal components present. The manufacturer has rated the inductance value based on the normal operating environment for this type inductor. *Students and instructors* should be aware that this inductor will exhibit "apparent" inductance values in our projects that are quite different from the manufacturer's rating. Due to the inductor being used under operating conditions different from those specified by the manufacturer, such factors as "incremental permeability," and so forth, enter into the results in terms of how much acting inductance the inductor "appears" to have. Also, the voltage dropped by the inductor is due to the inductor's impedance—not just its reactance. Add into this scenario the tolerance of component values, variances in calibration of signal sources, test equipment, and so on, and it is obvious that results will vary from those that would result if the inductor were truly acting at the manufacturer-rated value and all the components and test equipment were perfectly calibrated.

2. For the *higher-frequency circuit* option, there are several important things to keep in mind. Be aware that the inductor called for in the circuits is a 100-mH inductor. Typically, the tolerance on these inductors is large; therefore, they may act like inductances from 80 mH to 100 mH+, even though rated at 100 mH.

   Most handheld DMMs have frequency limitations in terms of measuring ac voltages and currents. In many cases, the highest frequencies they should be used to measure are signals up to about 1 kHz. For other, more expensive, true RMS meters, the specifications may allow for voltage and current measurements up to about 10 kHz.

   Since a number of the higher-frequency circuit options for our projects ask for frequency runs as high as 5 kHz to 10 kHz, the measurements must be made with a DMM (bench-type or otherwise) rated to measure ac up to at least 10 kHz. Ideally, the meter should be a *True RMS* reading instrument, although the more common *averaging type* reading instrument will probably work as long as its frequency rating is sufficient.

   Alternatively, a scope can be used for measurements. However, great caution must be used to make sure that the signal source ground and the scope ground(s) are connected to the same end of the component across which voltage measurement is being made. If they are not, the two ground locations can short out a component or a portion of the circuit under test! In some cases, this may mean changing positions of components in the circuit for each measurement; in other cases, this may not be necessary.

3. Students should be careful to set the signal sources to the correct frequencies and voltage levels for changes in the circuit conditions they are examining. Normally, it will be necessary to check and reset the voltage level whenever source frequency settings, circuit components, or conditions are changed.

Inductance
Total Inductance in Series and Parallel

PROJECT 40

Name: _____   Date: _____

FIGURE 40-1

Lower-Frequency Circuit   Higher-Frequency Circuit

PROJECT PURPOSE To demonstrate, through circuit connections and measurements, that inductances in series add like resistances in series and that inductances in parallel are analyzed in the same fashion as resistances in parallel are analyzed.

PARTS NEEDED
- ☐ DMM
- ☐ Low-voltage ac source (function generator)
- ☐ CIS
- ☐ Inductors, 1.5 H, 95 Ω, or approximate (2)
- ☐ 100 mH (2)
- ☐ Resistor, 1 kΩ

SPECIAL NOTE:

For this project, we will observe the property of inductance to oppose a change in current by applying a continuously changing ac voltage to the circuit and noting the current-limiting effects. The higher the inductance, or L, the higher the opposition to ac current. By noting the ac circuit current with a single inductor, then two inductors in series, then two inductors in parallel, we will illustrate the effects on total inductance of connecting two coils in series and in parallel.

For convenience, the voltage drop across a 1-kΩ resistor will be used as a current indicator. Since $I = V/R$, the number of volts divided by 1 kΩ automatically yields I in mA (e.g., 10 volts across a 1-kΩ resistor indicates 10 mA through the resistor, and so on).

PROCEDURE

1. Measure the dc resistance of the two inductors that will be used for this project. Also, set the ac source voltage that will be used to 3 V_{rms}.

   ⚠ OBSERVATION

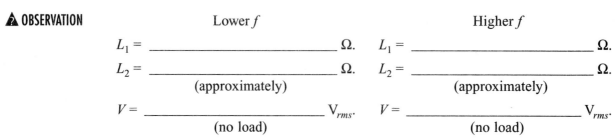

2. Connect the initial circuit as shown in Figure 40-1.

   ⚠ CONCLUSION If V_A were dc, what would be the current through this circuit? _____ mA.

3. Measure V_{R_1} (voltage drop across R_1) and calculate the ac current.

   ⚠ OBSERVATION

	Lower f	Higher f
$V_{R_1} =$	_____ V.	_____ V.
$I =$	_____ mA.	_____ mA.

   ⚠ CONCLUSION The back emf produced by the _____ is limiting the current to a lower value than it would be if dc were applied to the circuit.

4. Insert the second L (L_2) in series with the circuit. With V_A again set to 3 V_{rms}, measure V_{R_1} and calculate the ac current.

   ⚠ OBSERVATION

	Lower f	Higher f
$V_{R_1} =$	_____ V.	_____ V.
$I =$	_____ mA.	_____ mA.

   ⚠ CONCLUSION Since the current was lower with the two inductors in series, the L total is obviously (*more, less*) _____ than with one inductor. We conclude that inductors in series add like resistors in (*series, parallel*) _____.

5. Change the circuit so L_2 is in parallel with L_1. Measure V_{R_1} and calculate the ac current.

   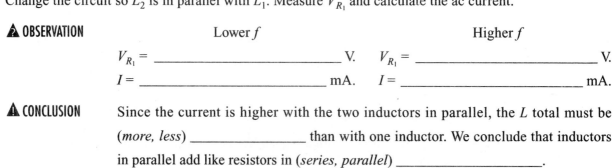

   ⚠ CONCLUSION Since the current is higher with the two inductors in parallel, the L total must be (*more, less*) _____ than with one inductor. We conclude that inductors in parallel add like resistors in (*series, parallel*) _____.

Story Behind the Numbers
Inductance

Name: _____   Date: _____

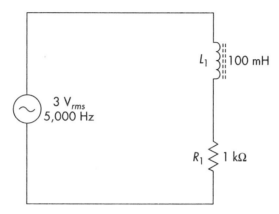

NOTE ▶ Prior to performing this project and the upcoming projects that will use inductors, be sure to read the **Special Notes to Students and Instructors** at the beginning of Part 9. For this project, you will use the higher-frequency circuit option; therefore, pay attention to the cautions regarding DMM measuring limitations (frequency limits). If you are using a DMM, be sure it is rated to properly read voltages at the frequencies required in this project. If a scope is to be used, observe the critical ground connection precautions necessary to prevent shorting out components.

GENERAL INFORMATION For this project, you will observe the property of inductance to oppose a change in current. You will do this by applying a continuously changing ac voltage to the circuits and noting the current-limiting effects. The higher the inductance (L), the higher the opposition to ac current. By noting the ac circuit current with a single inductor, two inductors in series, and two inductors in parallel, you will be able to illustrate the effects of total inductance on circuit current when the ac applied voltage is the same in all cases.

Procedure

1. Measure the dc resistances of the inductors that will be used for this project and record these values in the appropriate locations in the Data Table.

2. Connect the initial one-inductor circuit, as shown in the circuit diagram.

   NOTE ▶ For this project, we are using the higher-frequency circuit and circuit parameters setup.

3. Measure the voltage drop across the 1-kΩ resistor and record this value in the appropriate location in the Data Table.

4. Use the measured voltage value and the color-coded value of R_1 to calculate the circuit current value. Record in the Data Table, as appropriate.

5. Turn off the signal source and modify the circuit by adding a second inductor *in series* with the original inductor.

6. Turn on the signal source and set V_A to 3 V_{rms} again. Measure the voltage across the 1-kΩ resistor again. Record this value in the Data Table, as appropriate.

7. Calculate the circuit current for the circuit with the two inductors in series and record this value in the Data Table.

8. Turn off the signal source and modify the circuit by moving the second inductor so that it is now connected *in parallel* with the original circuit inductor.

9. Once again, measure the voltage drop across the 1-kΩ resistor and calculate the circuit current for the circuit having the two inductors in parallel. Record these parameters in the Data Table, as appropriate.

10. After completing the Data Table, answer the Analysis Questions and create the brief Technical Lab Report to complete the project.

Analysis Questions

NOTE ➤ Answers to these Analysis Questions should be clearly numbered and documented on separate sheets of paper with your name and the date at the top of each page. These answer sheets are to be turned in with the rest of the project documentation, as appropriate.

1. Explain why the current through the initial circuit was lower than it would have been if the applied voltage had been dc, rather than ac.

2. Explain why the current through the circuit with the two inductors in series was lower than when there was only one inductor in the circuit.

3. Was the circuit current with the two inductors *in series* equal to exactly one-half the value of the current in the single-inductor circuit? Should it have been? If not, explain what circuit parameters could account for this not being true.

4. Was the circuit current with the two inductors *in parallel* equal to exactly double the value of current in the single-inductor circuit? What component(s) and circuit factors might cause it to be less than double in this case?

5. If we had used *perfect* inductances (with no inherent coil resistance) and had not added an external resistance in series with the circuits, would the current value changes have been theoretically halved and doubled, as expressed in the previous steps?

6. List the mathematical expressions for total inductance of inductances in series and of inductances in parallel.

Data Table

Component and Parameter Identifiers	Higher f (1 inductor) $V_A = 3 V_{rms}$; $f = 5,000$ Hz Parameter Values
L_1 rated (H)	
L_1 measured R (Ω)	
R_1 color code (Ω)	
If V = dc, I would be:	
V_{R_1} measured (V_{rms})	
I (ac) calculated	
Component and Parameter Identifiers	**Higher f (2 series inductors) $V_A = 3 V_{rms}$; $f = 5,000$ Hz Parameter Values**
L_1 rated (H)	
L_2 rated (H)	
L_1 measured R (Ω)	
L_2 measured R (Ω)	
R_1 color code (Ω)	
V_{R_1} (ac) measured	
I (ac) calculated	
Component and Parameter Identifiers	**Higher f (2 parallel inductors) $V_A = 3 V_{rms}$; $f = 5,000$ Hz Parameter Values**
L_1 rated (H)	
L_2 rated (H)	
L_1 measured R (Ω)	
L_2 measured R (Ω)	
R_1 color code (Ω)	
V_{R_1} (ac) measured	
I (ac) calculated	

Technical Lab Report

Write a brief technical lab report summarizing the technical facts learned from this project. The report should be organized to provide the following:

1. An introductory paragraph describing the type of circuit being analyzed and the key parameters that will be discussed relating to this circuit.

2. A section describing the most important characteristics of this type of circuit that were shown via the collected data in the tables and graphs.

3. Any special facts or characteristics about this type of circuit that were highlighted in answering the Analysis Questions.

4. A practical example of how the information learned in this project might help you in operating, troubleshooting, error analysis, or adjusting a circuit of this type in your home setting, in your training program setting, or in a job setting in the real world.

5. A summary statement listing the most positive aspects of the project and any parts of the project that were difficult because of equipment problems or unclear instructions. Include areas that might be improved.

Summary
Inductance

Name: _____ Date: _____

Complete the following review questions, indicating the appropriate response by placing a check in the box next to the correct answer.

1. Inductance is that property in an electrical circuit that opposes
   - ☐ a change in voltage
   - ☐ a change in current
   - ☐ a change in resistance

2. In an ac circuit, if L is increased, the circuit current will
   - ☐ increase
   - ☐ remain the same
   - ☐ decrease

3. If two equal inductances are connected in series, total inductance will be
   - ☐ two times that of one
   - ☐ neither of these
   - ☐ one-half that of one

4. If two equal inductances are connected in parallel, total inductance will be
   - ☐ two times that of one
   - ☐ neither of these
   - ☐ one-half that of one

INDUCTIVE REACTANCE IN AC

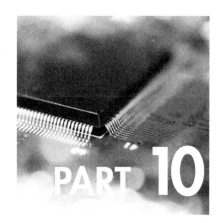

PART 10

Objectives

You will connect several ac inductive circuits and make measurements and observations regarding their important electrical characteristics.

In completing these projects, you will connect circuits, make measurements, perform calculations, draw conclusions, and be able to answer questions about the following items related to inductive reactance:

- Relationship of L, induced voltage, and inductive reactance
- Relationship of frequency to inductive reactance
- The X_L formula
- Solving for L when X_L and frequency are known

Project/Topic Correlation Information

PROJECT	TEXT CHAPTER	SECTION	RELATED TEXT TOPIC(S)
41 Induced Voltage	13	13-2	Review of Faraday's and Lenz's Laws
		13-3	Self-Inductance
42 Relationship of X_L to L and Frequency	14	14-4	Relationship of X_L to Inductance Value
		14-5	Relationship of X_L to Frequency of AC

SPECIAL NOTE:

For this project, and all the following projects that require *setting* or *measuring* ac voltages (or currents), assume that the rms values are desired, unless specified otherwise.

Inductive Reactance in AC
Induced Voltage

PROJECT 41

Name: _____ Date: _____

FIGURE 41-1

Lower-Frequency Circuit: V_A = 3 V_{rms}, 100 Hz; L = 1.5 H (rating); R_1 = 1 kΩ

Higher-Frequency Circuit: V_A = 3.0 V_{rms}, 5 kHz; L_1 = 100 mH; R_1 = 1 kΩ

PROJECT PURPOSE To illustrate the circuit effects of inductor counter-emf by noting the difference in circuit current when dc is applied to the inductor circuit, and again when ac is applied.

PARTS NEEDED
- ☐ DMM
- ☐ VVPS (dc)
- ☐ Low-voltage ac source (function generator, approximately 7 VAC)
- ☐ CIS
- ☐ Inductor, 1.5 H, 95 Ω (or approximate)
- ☐ 100 mH
- ☐ Resistor 1 kΩ

SPECIAL NOTE:

The methods used during this project are used to illustrate the concept of back emf and are not a precise scientific method of measuring or calculating exact values of back emf. Also remember that the ac equivalent of a given dc value is the "effective" (rms) value of ac.

PROCEDURE

1. Connect the initial circuit as shown in Figure 41-1.

2. Measure the ac voltage source that will be used for this demonstration. Next, connect the VVPS to the circuit and adjust the dc input V to a value that matches the ac voltage you will be using in a later step.

   ⚠ OBSERVATION AC source equals _____ V_{rms}.

   DC source connected to circuit set to _____ V.

3. Measure V_{R_1} (dc voltage across R_1) and calculate circuit dc current.

   ⚠ OBSERVATION (Using lower f circuit components)   (Using higher f circuit components)

   I (dc) = _____ mA. I (dc) = _____ mA.

   ⚠ CONCLUSION The circuit current is limited only by the dc resistance of R_1 and _____.

4. Remove the dc source, connect the 3 V_{rms} ac source and measure V_1. Now calculate the circuit ac current.

   ⚠ OBSERVATION Lower f Higher f

   I (ac) = _____ mA. I (ac) = _____ mA.

   ⚠ CONCLUSION The inductor developed a back emf that opposes the changing current. Discount the small dc resistance of the coil. The current that flowed with dc applied was equivalent to having a V_A of _____ V across a circuit resistance of 1 kΩ. When the same value of ac voltage was applied to the circuit, a current flowed that would be equivalent to applying only _____ V across the circuit resistance. The effect simulated was as if there must be a back emf of approximately _____ V.

   NOTE ▶ This disregards any R in the circuit and assumes that the total limiting effect on the current is due to back emf.

Inductive Reactance in AC
Relationship of X_L to L and Frequency

PROJECT 42

Name: _____ Date: _____

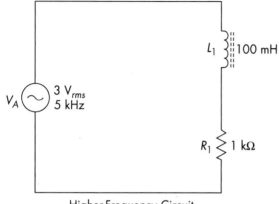

Higher-Frequency Circuit

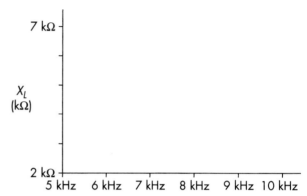

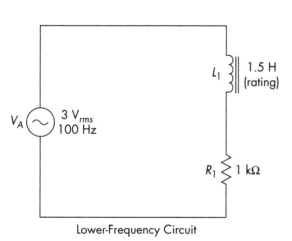

Lower-Frequency Circuit

FIGURE 42-1

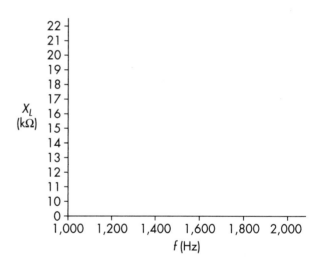

FIGURE 42-2 X_L versus f (sample graph coordinates)

NOTE If using a different value of L than shown, renumber the y-axis of the graph, as appropriate. Use separate graph paper for drawing these graphs.

PROJECT PURPOSE To observe the relationship of X_L to L by changing L_T, and to frequency by applying various frequencies to an inductor circuit, and by measuring the changing circuit parameters.

PARTS NEEDED
- ☐ DMM
- ☐ Function generator or audio oscillator
- ☐ CIS
- ☐ Inductor, 1.5 H, 95 Ω (2) (or approximate)
- ☐ 100 mH (2)
- ☐ Resistor 1 kΩ

197

SPECIAL REMINDER:

The inductors used in the lower-frequency circuit of this project, *and a number of projects following this one*, are iron-core, filter-choke-type inductors. Under normal operating conditions, this inductor has dc current as well as ac signal components present. The manufacturer has rated the inductance value based on the normal operating environment for this type of inductor. *Students and instructors* should be aware that this inductor will exhibit "apparent" inductance values quite different from the manufacturer's rating under the varying operating conditions in our projects. Because the inductor is used under different conditions from those specified by the manufacturer, such factors as "incremental permeability" enter into the results in terms of how much acting inductance the inductor appears to have. Also, the voltage dropped by the inductor is due to the inductor's impedance—not just its reactance. Add into this scenario the tolerances of component values, variances in calibration of signal sources, test equipment, and so on, and it is obvious that results will vary from those that would result if the inductor were acting at the rated value.

PROCEDURE

1. Connect the initial circuit as shown in Figure 42-1.

2. Set the function generator to a frequency of 100 Hz for the lower f circuit or to 5,000 Hz for the higher f circuit and set V_A from the source to 3 V. Measure V_{R_1} and calculate the circuit current. Next, measure V_L and calculate X_L by Ohm's law ($X_L = V_L/I$).

   ⚠ OBSERVATION

   Lower f

   $V_A =$ _____ V.
   $V_{R_1} =$ _____ V.
   $I =$ _____ mA.
   $V_L =$ _____ V.
   $X_L =$ _____ Ω.

   Higher f

   $V_A =$ _____ V.
   $V_{R_1} =$ _____ V.
   $I =$ _____ mA.
   $V_L =$ _____ V.
   $X_L =$ _____ Ω.

   ⚠ CONCLUSION Since this is a simple series circuit, the current through the inductor is the same as the current through the _____.

3. Insert a second inductor (the same type as L_1). You should now have a series circuit of L_1, L_2, and R_1. With the same frequency and V_A as step 2 above, measure V_{R_1}, calculate I, measure the voltage across the total inductance of L_1 and L_2, then calculate X_L total.

   ⚠ OBSERVATION

   Lower f

   $V_A =$ _____ V.
   $V_{R_1} =$ _____ V.
   $I =$ _____ mA.
   $V_{L_T} =$ _____ V.
   $X_{L_T} =$ _____ Ω.

   Higher f

   $V_A =$ _____ V.
   $V_{R_1} =$ _____ V.
   $I =$ _____ mA.
   $V_{L_T} =$ _____ V.
   $X_{L_T} =$ _____ Ω.

▲ CONCLUSION Connecting a second inductor of nearly equal value as L_1 in series with L_2 caused the total inductance to approximately (*double, halve*) _____. In analyzing the results of step 2 and comparing with this step, we conclude that doubling total inductance caused the total inductive reactance (X_L) to approximately (*double, halve*) _____. We therefore conclude that inductive reactance (X_L) is (*directly, inversely*) _____ proportional to inductance (L). Increasing L causes X_L to (*increase, decrease*) _____ at any given frequency.

4. Remove L_2 and replace with a jumper. For the lower f circuit, change the input frequency to 1,000 Hz and keep V_A at 3 volts. For the higher f circuit, change the input frequency to 10,000 Hz, keeping V_A at 3 volts. Measure V_{R_1}, calculate I, measure V_L, and calculate X_L.

▲ OBSERVATION

Lower f		Higher f	
V_A = _____	V.	V_A = _____	V.
V_{R_1} = _____	V.	V_{R_1} = _____	V.
I = _____	mA.	I = _____	mA.
V_{L_1} = _____	V.	V_{L_1} = _____	V.
X_L = _____	kΩ.	X_L = _____	kΩ.

▲ CONCLUSION Increasing the frequency caused the inductive reactance (X_L) to (*increase, decrease*) _____. If this had been a perfect inductive circuit, would X_L have increased the same number of times as frequency was increased? _____.

We conclude that inductive reactance is (*inversely, directly*) _____ proportional to f (frequency). If the frequency were decreased to one-half its original value, then theoretically the X_L would (*increase, decrease*) _____ to (*double, one-half*) _____ its original value.

5. Make a graphic plot of X_L versus f from 1,000 Hz to 2,000 Hz for the lower f circuit or from 5,000 to 10,000 Hz for the higher f circuit. Use graph coordinates similar to those shown in Figure 42-2. Make graphs on separate graph paper.

NOTE ▶ Calculate X_L and "apparent L" values for each 200-Hz or each 1,000-Hz change, as appropriate. See optional steps 9 and 10 for an alternative approach.

▲ OBSERVATION

Lower f			Higher f		
Apparent L values			Apparent L values		
L = _____	X_L at 1,000 Hz = _____		L = _____	@ 5 kHz = _____	kΩ.
L = _____	X_L at 1,200 Hz = _____		L = _____	@ 6 kHz = _____	kΩ.
L = _____	X_L at 1,400 Hz = _____		L = _____	@ 7 kHz = _____	kΩ.
L = _____	X_L at 1,600 Hz = _____		L = _____	@ 8 kHz = _____	kΩ.
L = _____	X_L at 1,800 Hz = _____		L = _____	@ 9 kHz = _____	kΩ.
L = _____	X_L at 2,000 Hz = _____		L = _____	@ 10 kHz = _____	kΩ.

▲ CONCLUSION Did X_L act like it is directly related to f? _____.

Optional Steps

6. Use the appropriate version of the $X_L = 2\pi fL$ formula to find the apparent L of the inductor for the operating conditions used in step 2.

   NOTE ➤ Use the X_L value found by Ohm's law in step 2 when solving for L.

   ⚠ OBSERVATION
   Lower f
   Calculated apparent
   $L = $ _____ H.

   Higher f
   Calculated apparent
   $L = $ _____ mH.

   ⚠ CONCLUSION Is the calculated (apparent) inductance different from the manufacturer's rated value for this inductor? _____. What do you think the inductance value would appear to be if the inductor were operated under the conditions used by the manufacturer when rating the inductor? _____ H.

7. Use the appropriate version of the $X_L = 2\pi fL$ formula to find the apparent L_T of the series inductors for the operating conditions used in step 3.

   NOTE ➤ Use the X_L value found by Ohm's law in step 3 when solving for L_T.

   ⚠ OBSERVATION
   Lower f
   Calculated apparent
   $L_T = $ _____ H.

   Higher f
   Calculated apparent
   $L_T = $ _____ mH.

   ⚠ CONCLUSION Is the calculated (apparent) total inductance different from the manufacturer's rated value for these inductors? _____. What do you think the total inductance value would appear to be if the inductors were operated under the conditions used by the manufacturer when rating these inductors? _____ H.

8. Use the appropriate version of the $X_L = 2\pi fL$ formula to find the apparent L of the inductor for the operating conditions used in step 4.

   NOTE ➤ Use the X_L value found by Ohm's law in step 4 when solving for L.

   ⚠ OBSERVATION
   Lower f
   Calculated apparent
   $L = $ _____ H.

   Higher f
   Calculated apparent
   $L = $ _____ mH.

   ⚠ CONCLUSION Is the calculated (apparent) inductance different from the manufacturer's rated value for this inductor, in this case? _____. Is it different under step 4's operating conditions than it was for step 2? _____. If so, what might account for the difference?

   _____.

9. Use Excel (or another) spreadsheet and set up a chart similar to the one shown below for charting the data collected in step 5.

	A	B	C	D
	PROJECT: RELATIONSHIP OF X_L TO FREQUENCY *(Data derived from Project 42, step 5 data)*			
1	Frequency	Mfg Rated	Calc X_L	Calc X_L
2	Setting (Hz)	L in H	X_L formula	From Measurements
3	1st Freq (Hz)	L rating (H)	= 6.28*A3*B3	= V_L/I
4	2nd Freq (Hz)	L rating (H)	= 6.28*A4*B4	= V_L/I
5	(etc.)	(etc.)	(etc.)	(etc.)
6	(etc.)	(etc.)	(etc.)	(etc.)
7	(etc.)	(etc.)	(etc.)	(etc.)

10. Use the spreadsheet chart graphing capability to create a line graph showing frequency as the x-axis and X_L as the y-axis data. Show a line graph for both the X_L formula results (column C) and the X_L from measurements results (column D) on the same graph. Appropriately label the graph using the titling and labeling capabilities of the spreadsheet program.

Story Behind the Numbers
Inductive Reactance in AC

Name: _____   Date: _____

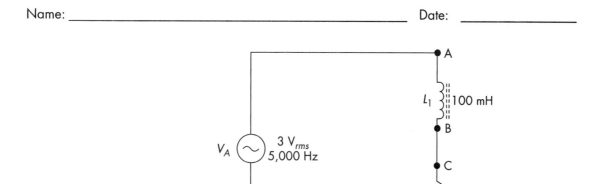

NOTE ➤ Prior to performing this project and the upcoming projects that will use inductors, be sure to read the **Special Notes to Students and Instructors** at the beginning of Part 9. For this project, you will use the higher-frequency circuit option; therefore, pay attention to the cautions regarding DMM measuring limitations (frequency limits). If you are using a DMM, be sure it is rated to properly read voltages at the frequencies required in this project. If a scope is to be used, observe the critical ground connection precautions necessary to prevent shorting out components.

GENERAL INFORMATION In this project, you will first look at how inductance value affects circuit inductive reactance. Then, you will change circuit source frequency over a range of frequencies to see how frequency affects inductive reactance for a given circuit setup.

Procedure

1. Connect the initial circuit as shown.

2. With 3 volts applied voltage and the signal source set at 5,000 Hz, measure V_{R_1} and calculate the circuit current. Record this data in the appropriate locations in the Data Table.

3. Measure V_L and calculate X_L, using Ohm's law (V_L/I). Record this value in the Data Table, as appropriate.

4. Turn off the signal source. Remove the circuit jumper between circuit points B and C. Insert a second 100-mH inductor at circuit points B and C, so as to have two 100-mH inductors in series.

5. Turn on the signal source, making sure that the circuit applied voltage is set at 3 volts and the source frequency is still set at 5,000 Hz.

6. Measure V_{R_1}, calculate circuit current, and record data in the Data Table.

7. Measure V_{L_T} (voltage across both inductors in series) and calculate X_{L_T} using Ohm's law. Record data, as appropriate, in the Data Table.

8. Turn off the signal source. Return the circuit configuration to the original one-inductor "initial" circuit setup using the inductor whose value is rated at 100 mH.

9. Now, make a "frequency run" for this circuit as follows:
   a. Set the signal source to a frequency of 5,000 Hz and its output voltage to 3 V.
   b. Use the techniques you used earlier to measure V_{R_1}, calculate circuit I, measure V_L, and calculate X_L.
   c. Measure and record data in Data Table 2, as indicated, to show inductive reactance for each new frequency setting from 5,000 to 10,000 Hz in 1,000-Hz steps. That is, record values you determined at 5,000 Hz, 6,000 Hz, 7,000 Hz, and so on.

10. Create a line graph showing X_L versus frequency (f) over the frequency run. Use f settings along the **x-axis**, and let the **y-axis** represent the X_L values.

11. After completing the Data Table and creating the graph, answer the Analysis Questions and produce the brief Technical Lab Report to complete the project.

    NOTE ▶ If your instructor wants you to perform measurements and analysis on both the higher-frequency circuit(s) and the lower-frequency circuit(s), perform steps 1–10 a second time, using the correct components and frequency settings. Record data, as appropriate, in the Data Table, and create graphs for both types of circuits.

Analysis Questions

NOTE Answers to these Analysis Questions should be clearly numbered and documented on separate sheets of paper with your name and the date at the top of each page. These answer sheets are to be turned in with the rest of the project documentation, as appropriate.

1. Why was the circuit inductive reactance greater with the two inductors in series than it was with the single inductor?

2. From the data in Data Table 1, use the appropriate version of the $X_L = 2\pi f L$ formula and calculate the "apparent" *inductance* of the single inductor used in the initial circuit. Show your work.

3. From the data in Data Table 1, use the appropriate version of the $X_L = 2\pi f L$ formula to find the *total inductance* of the two inductors in series. Show your work.

4. Was the inductance of the two-inductor circuit approximately two times the inductance of the one-inductor circuit? Was the total inductive reactance of the two-inductor circuit approximately two times the inductive reactance

Data Table 1

Component and Parameter Identifiers	Higher f (1 inductor) $V_A = 3\ V_{rms}$; @ $f = 5{,}000$ Hz Parameter Values
L_1 rated (H)	0.1
R_1 color code (Ω)	1,000
V_{R_1} measured (V_{rms})	
I (ac) calculated (mA)	
V_L measured (V_{rms})	
X_L calculated (kΩ)	
Component and Parameter Identifiers	**Higher f (2 series inductors) $V_A = 3\ V_{rms}$; @ $f = 5{,}000$ Hz Parameter Values**
L_1 rated (H)	0.1
L_2 rated (H)	0.1
R_1 color code (Ω)	1,000
V_{R_1} measured (V_{rms})	
I (ac) calculated (mA)	
V_{L_T} measured (V_{rms})	
X_{L_T} calculated (kΩ)	

of the single-inductor circuit having the same circuit voltage applied and frequency parameters?

5. From your answers in question 4, what can you describe about the relationship of inductive reactance to inductance value for given circuit inductor(s) at a specific voltage applied and source frequency?

6. From the data in Data Table 2 (the frequency run data), what relationship can you describe about the value of inductive reactance as it relates to frequency for a given circuit configuration?

7. Use the appropriate version of the $X_L = 2\pi fL$ formula and calculate the "apparent" *inductance* of the single inductor used for the frequency run at the lowest frequency used in the run and at the highest frequency in the run. Show your work.

8. Were the apparent inductance values the same at the lower-frequency end and the higher-frequency end of the run? If not, explain some factors that may have caused this.

9. Describe why your calculated apparent inductance values may differ somewhat from the manufacturer's rated value for these inductors.

Data Table 2

Component and Parameter Identifiers	Higher f	Circuit	(1 inductor)	$V_A = 3\ V_{rms}$	@ f = 5,000–10,000 Hz	
	5,000 Hz	6,000 Hz	7,000 Hz	8,000 Hz	9,000 Hz	10,000 Hz
V_{R_1} measured (V_{rms})						
I (ac) calculated (mA)						
V_L measured (V_{rms})						
X_L calculated (kΩ)						

Technical Lab Report

Write a brief technical lab report summarizing the technical facts learned from this project. The report should be organized to provide the following:

1. An introductory paragraph describing the type of circuit being analyzed and the key parameters that will be discussed relating to this circuit.

2. A section describing the most important characteristics of this type of circuit that were shown via the collected data in the tables and graphs.

3. Any special facts or characteristics about this type of circuit that were highlighted in answering the Analysis Questions.

4. A practical example of how the information learned in this project might help you in operating, troubleshooting, error analysis, or adjusting a circuit of this type in your home setting, in your training program setting, or in a job setting in the real world.

5. A summary statement listing the most positive aspects of the project and any parts of the project that were difficult because of equipment problems or unclear instructions. Include areas that might be improved.

Summary
Inductive Reactance in AC

Name: _____ Date: _____

Complete the following review questions, indicating the appropriate response by placing a check in the box next to the correct answer.

1. The induced voltage and X_L of a coil are directly proportional to
   - ☐ L and R
   - ☐ L/R
   - ☐ L and f
   - ☐ R and f
   - ☐ none of these

2. If frequency is doubled and L is halved, the net resultant X_L will
   - ☐ double
   - ☐ halve
   - ☐ quadruple
   - ☐ remain the same
   - ☐ none of these

3. If two equal inductors that were in series are now parallel connected, the resulting total X_L compared to the original circuit will be
   - ☐ two times greater
   - ☐ one-half as great
   - ☐ four times greater
   - ☐ one-fourth as great

4. As f increases, the rate of change of current
   - ☐ increases
   - ☐ decreases
   - ☐ remains the same

5. The X_L formula shows that inductive reactance is
   - ☐ directly proportional to L and inversely proportional to f
   - ☐ inversely proportional to L and directly proportional to f
   - ☐ neither of these

6. To solve for L when X_L and frequency are known, use the formula
   - ☐ $2\pi f/X_L$
   - ☐ $2\pi X_L/f$
   - ☐ $X_L/2\pi f$
   - ☐ none of these

7. The unit of X_L is the
   - ☐ back emf
   - ☐ ohm
   - ☐ ampere
   - ☐ volt
   - ☐ none of these

8. The opposition that an inductor shows to ac is
   - ☐ purely inductive
   - ☐ purely resistive
   - ☐ a combination of resistance and inductive reactance
   - ☐ none of these

9. The amount of inductive reactance that a given coil exhibits is directly related to
   - ☐ the amount of current
   - ☐ the applied voltage
   - ☐ neither of these

10. If frequency is tripled and inductance halved, the resultant X_L will be
    - ☐ two-thirds of the original
    - ☐ three-halves of the original
    - ☐ six times the original
    - ☐ one-sixth of the original

RL CIRCUITS IN AC

PART 11

Objectives

You will connect several ac *RL* circuits and make measurements and observations regarding their important electrical characteristics.

In completing these projects, you will connect circuits, make measurements, perform calculations, draw conclusions, and be able to answer questions about the following items related to *RL* circuits:

- Relationship of R and X_L to circuit phase angle
- Relationships of current and voltage for an inductor and for a resistor
- Circuit impedance
- Simple vector diagram(s)
- Reference vector(s)

Project/Topic Correlation Information

PROJECT	TEXT CHAPTER	SECTION	RELATED TEXT TOPIC(S)
43 *V, I, R, Z,* and θ Relationships in a Series *RL* Circuit	15	15-4	Fundamental Analysis of Series *RL* Circuits
44 *V, I, R, Z,* and θ Relationships in a Parallel *RL* Circuit	15	15-5	Fundamental Analysis of Parallel *RL* Circuits

RL Circuits in AC
V, I, R, Z, and θ Relationships in a Series RL Circuit

PROJECT 43

Name: _____ Date: _____

FIGURE 43-1

Lower-Frequency Circuit: V_A = 3 V_{rms}, 500 Hz; L_1 = 1.5 H (rating); R_1 = 1 kΩ

Higher-Frequency Circuit: V_A = 3 V_{rms}, 2 kHz; L_1 = 100 mH; R_1 = 1 kΩ

PROJECT PURPOSE To demonstrate the key electrical parameter relationships in a series RL circuit. To observe that due to out-of-phase elements, simple dc analysis techniques cannot be used to determine circuit parameters in ac circuits containing reactive components. Furthermore, to provide practice in using simple ac analysis techniques and in drawing ac circuit vector diagrams.

PARTS NEEDED
- ☐ DMM
- ☐ Function generator or audio oscillator
- ☐ Dual-trace oscilloscope
- ☐ CIS
- ☐ Inductors 1.5 H, 95 Ω
- ☐ 100 mH
- ☐ Resistor 1 kΩ

PROCEDURE

1. Connect the initial circuit as shown in Figure 43-1.

2. Set the frequency of the function generator to the frequency indicated for the circuit option you are using. Set the circuit input voltage to 3 volts. Measure V_A, V_{R_1}, and V_{L_1}.

⚠ OBSERVATION

Lower f
- V_A = _____ V.
- V_{R_1} = _____ V.
- V_{L_1} = _____ V.

Higher f
- V_A = _____ V.
- V_{R_1} = _____ V.
- V_{L_1} = _____ V.

▲ CONCLUSION   Does V_A equal the arithmetic sum of V_{R_1} and V_{L_1}? _____. We conclude that to find V_A we must vectorially (*add, subtract*) _____ V_{R_1} and V_{L_1}. We can also use the _____ theorem.

3. Calculate I_T from V_{R_1}/R_1. Calculate X_L from V_{L_1}/I. Calculate circuit total impedance from $Z = V_T/I_T$.

   ▲ OBSERVATION   Lower f Higher f

   I_T = _____ mA. I_T = _____ mA.
   X_L = _____ Ω. X_L = _____ Ω.
   Z = _____ Ω. Z = _____ Ω.

4. Determine the apparent value of L from the known frequency and X_L parameters.

   ▲ OBSERVATION   Lower f Higher f

   L_1 = _____ H. L_1 = _____ H.

5. Determine Z using the Pythagorean theorem (using the R and X_L parameters).

   ▲ OBSERVATION   Lower f Higher f

   Z = _____ Ω. Z = _____ Ω.

6. Draw an impedance diagram in the "Observation" section.

   ▲ OBSERVATION

7. Use trigonometry and determine the phase angle.

   ▲ OBSERVATION   Lower f Higher f

   θ = _____ degrees.   θ = _____ degrees.

   ▲ CONCLUSION   These conclusions are drawn for steps 3 through 7.

   Does Z equal the arithmetic sum of R_1 and X_L? _____. We again conclude that we must use the vector sum or the Pythagorean theorem used in analysis of right _____. From the observations of steps 2 and 3 we may conclude that V_{R_1} and V_{L_1} are (*in phase, not in phase*) _____. Since inductance-opposes a change in current, we may assume that V_{L_1} (*leads, lags*) _____ I_L by some angle. If L_1 were a perfect inductor, _____ would lead _____ by 90 degrees. Since the circuit is not composed of pure resistance in which the phase angle between V and $I =$ _____ degrees, nor pure inductance in which the phase angle between V and I equals _____ degrees, but rather a composite of both, we might expect the phase angle between V_T and I_T to be between _____ and _____ degrees. Further, the larger the X_L is compared to the circuit R, the more like a purely inductive circuit the results will be; thus the (*greater, lesser*) _____ will be the circuit phase angle. The converse is also true.

Optional Steps

8. Use the measured and calculated data in steps 2 and 3, and draw a *V-I* vector diagram in the "Observation" section.

   ⚠ OBSERVATION

9. Use trigonometry and determine the phase angle.

   ⚠ OBSERVATION   Lower f Higher f

   θ = _____ degrees.   θ = _____ degrees.

   ⚠ CONCLUSION Does the phase angle from the *V-I* vector diagram agree reasonably with the phase angle you determined from the Z diagram? _____.

10. Use a dual-trace oscilloscope and perform a phase comparison of V_A and circuit current (represented by the voltage across the resistor).

    ❗ CAUTION: Be sure the signal source ground and the scope ground(s) are connected to the same end of the resistor when making the measurements to prevent the two grounds from shorting out a portion of the circuit!

    Determine the phase difference between the two signals. (If possible, demonstrate your scope waveforms and calculations to your instructor.)

⚠ OBSERVATION θ determined by the scope phase comparison:

                        Lower f Higher f

                θ = _____ degrees.   θ = _____ degrees.

⚠ CONCLUSION Do the scope phase measurements and the phase angle calculations agree reasonably with your earlier findings? (Considering tolerances in components, source and scope frequency calibration tolerances, etc.) _____.

RL Circuits in AC
V, I, R, Z, and Θ Relationships in a Parallel RL Circuit

PROJECT 44

Name: _____ Date: _____

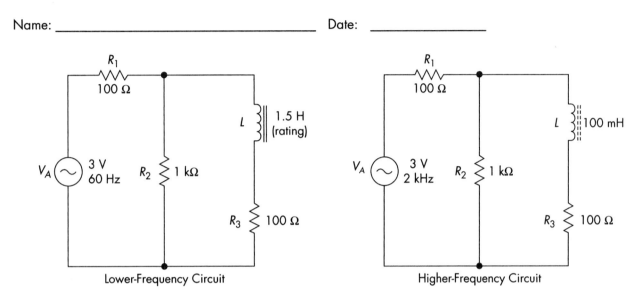

FIGURE 44-1

PROJECT PURPOSE To demonstrate the key electrical parameter relationships in a parallel *RL* circuit. To observe that due to out-of-phase elements, simple dc analysis techniques cannot be used to determine circuit parameters in ac circuits containing reactive components. Also, to provide practice in using simple ac analysis techniques and in drawing ac circuit vector diagrams.

PARTS NEEDED
- ☐ DMM
- ☐ Function generator or audio oscillator
- ☐ Low-voltage 60-Hz ac source
- ☐ CIS
- ☐ Inductors
  1.5 H, 95 Ω (or approximate)
- ☐ 100 mH
- ☐ Resistors
  100 Ω (2)
  1 kΩ (1)

SPECIAL NOTE:

It should be noted that, once again, we will be using the 100-Ω resistor in series with the main line as a circuit current indicator. The current equals ten times the voltage drop.

PROCEDURE

1. Connect the initial circuit as shown in Figure 44-1.

2. Set the source for 60 Hz, 3 V_{rms} for the lower f circuit and for 2,000 Hz, 3 V_{rms} for the higher f circuit. Measure the circuit voltages.

OBSERVATION

Lower f		Higher f	
$V_A =$	V.	$V_A =$	V.
$V_{R_1} =$	V.	$V_{R_1} =$	V.
$V_{R_2} =$	V.	$V_{R_2} =$	V.
$V_{R_3} =$	V.	$V_{R_3} =$	V.
$V_L =$	V.	$V_L =$	V.

CONCLUSION Do the voltages around any given closed loop add up by addition to V_A? _____.

From this we conclude that the circuit current(s) and voltage(s) are (*in phase, out of phase*) _____.

3. Calculate the total circuit current from V_{R_1}. Calculate the current through R_2 by Ohm's law. Calculate the current through L by using the voltage drop across R_3. Use Ohm's law to solve for I_L ($I_L = V_3/R_3$) and X_L ($X_L = V_L/I_L$).

OBSERVATION

Lower f		Higher f	
$I_T =$	mA.	$I_T =$	mA.
$I_{R_2} =$	mA.	$I_{R_2} =$	mA.
$I_L =$	mA.	$I_L =$	mA.
$X_L =$	Ω.	$X_L =$	Ω.

4. Use measured and calculated values of I_{R_2} and I_L. Apply the Pythagorean theorem formula and calculate I_T.

OBSERVATION

Lower f		Higher f	
I_T calculated =	mA.	I_T calculated =	mA.

5. Draw the appropriate *V-I* vector diagram in the "Observation" section.

OBSERVATION

6. Use trigonometry and determine the phase angle. (Neglect R_1 parameters and use only R_2 and L parameters.)

⚠ OBSERVATION

                Lower f Higher f

        $\theta =$ _____ degrees. $\theta =$ _____ degrees.

⚠ CONCLUSION   These conclusions are drawn for steps 3 through 6.

Does total current equal the arithmetic sum of the branch currents? _____. This is because the branch currents are _____ _____ _____. The current through R_2 is in phase with V_2. (*True* or *False*) _____. The current through the coil (*leads, lags*) _____ the voltage across the coil by close to _____ degrees. If the inductor were perfect, it would be exactly _____ degrees. Since the total circuit current is the vector resultant of the two branch currents, it would seem logical to assume the circuit total current would be (*leading, lagging*) _____ V_A by some angle between _____ and _____ degrees. Also note from our measurements and calculations that the total circuit impedance (Z) cannot be found by the product-over-the-sum method but is most easily solved by Ohm's law, where $Z = V_T/I_T$.

Optional Step

7. Use a dual-trace oscilloscope and perform a phase comparison of the current through the R_2 branch and the current through the inductor branch. Do this by letting the voltage across R_2 represent the current through R_2 and the voltage across R_3 represent the current through the inductor branch.

    ! CAUTION: Be sure the signal source ground and the scope ground(s) are connected to the same end (the bottom end) of the circuit network when making the measurements to prevent the grounds from shorting out a portion of the circuit!

Determine the phase difference between the two signals. (If possible, demonstrate your scope waveforms and calculations to your instructor.)

⚠ OBSERVATION   Measured θ between the resistor branch and the inductor branch as determined by the scope phase comparison:

                Lower f Higher f

        $\theta =$ _____ degrees. $\theta =$ _____ degrees.

⚠ CONCLUSION   Was the phase difference between the two branches reasonably close to 90°? _____. If not, what variables and factors might account for the difference? _____ _____ _____ _____.

Story Behind the Numbers
Section 1: RL (Series) Circuits in AC

Name: _____   Date: _____

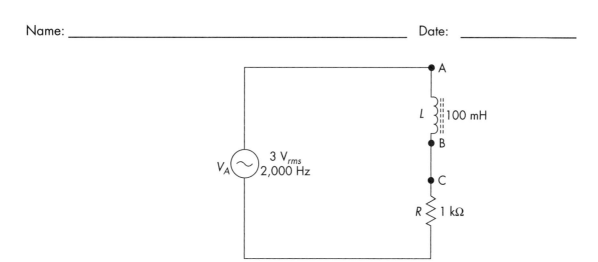

NOTE ▶ Prior to performing this project and the upcoming projects that will use inductors, be sure to read the **Special Notes to Students and Instructors** at the beginning of Part 9. For this project, you will use the higher-frequency circuit option; therefore, pay attention to the cautions regarding DMM measuring limitations (frequency limits). If you are using a DMM, be sure it is rated to properly read voltages at the frequencies required in this project. If a scope is to be used, observe the critical ground connection precautions necessary to prevent shorting out components.

Procedure

1. Connect the circuit as shown.

2. Set the frequency of the function generator to 2,000 Hz and its output to 3 V.

3. Measure V_A, V_R, and V_L and record the values in the Data Table where indicated.

4. From the measured parameter values, calculate the circuit total current (V_R/R), the value of X_L (V_L/I), and the circuit total impedance (V_T/I_T) and record these values in the Data Table where indicated.

5. Use the Pythagorean theorem and find the value of circuit Z using the rated value of R and the calculated value of X_L. Record this value in the Data Table where indicated.

6. Use the appropriate version of the X_L formula and determine the "apparent" value of L from the known frequency and X_L parameters. Record this value in the Data Table.

7. Use a dual-trace oscilloscope and perform a phase comparison of V_A and circuit current (represented by the voltage across the resistor).

! CAUTION: Be sure the signal source ground and the scope ground(s) are connected to the same end of the resistor when making the measurements to prevent the two grounds from shorting out a portion of the circuit!

Determine the phase difference between the two signals. If possible, demonstrate your scope pattern to your instructor.

8. After completing the Data Table, answer the Analysis Questions and develop a brief Technical Lab Report to complete the project.

Data Table

Component and Parameter I.D.'s	Measured Values	Calculated Values
V_A (V)		—
V_R (V)		—
V_L (V)		—
I_T (mA)	—	
X_L (kΩ)	—	
Z, Ohm's law (kΩ)	—	
Z, Pythagorean theorem (kΩ)	—	
Calculated apparent L (mH)	—	

Analysis Questions

NOTE Answers to these Analysis Questions should be clearly numbered and documented on separate sheets of paper with your name and the date at the top of each page. These answer sheets are to be turned in with the rest of the project documentation, as appropriate.

1. Review any data you have collected and determine if V_A equals the arithmetic sum of the values of V_R and V_L. Explain your finding.

2. Does the circuit Z equal the sum of the values of R and X_L? Explain your finding.

3. Does the value you found for the "apparent" L match the manufacturer's rating for the inductor used? Is it reasonably close? List several factors that may cause differences.

4. Use the data you have collected to draw an impedance diagram for the circuit conditions you used in this project. Label all parts of the diagram appropriately.

5. Use trigonometry to find the phase angle represented in your impedance diagram.

6. Use the data you have collected and draw a *V-I* vector diagram. Label all parts of the diagram appropriately.

7. Use trigonometry to find the phase angle represented in your *V-I* vector diagram.

8. Record the computation you used for determining the phase angle from your dual-trace scope pattern.

9. Define and list the changes that would occur (in general terms) for the various circuit parameters if the source frequency were doubled, while keeping the V_A at 3 volts. (Define each specific parameter in terms of whether it would increase, decrease, or remain the same.)

Technical Lab Report

Write a brief technical lab report summarizing the technical facts learned from this project. The report should be organized to provide the following:

1. An introductory paragraph describing the type of circuit being analyzed and the key parameters that will be discussed relating to this circuit.

2. A section describing the most important characteristics of this type of circuit that were shown via the collected data in the tables and graphs.

3. Any special facts or characteristics about this type of circuit that were highlighted in answering the Analysis Questions.

4. A practical example of how the information learned in this project might help you in operating, troubleshooting, error analysis, or adjusting a circuit of this type in your home setting, in your training program setting, or in a job setting in the real world.

5. A summary statement listing the most positive aspects of the project and any parts of the project that were difficult because of equipment problems or unclear instructions. Include areas that might be improved.

Story Behind the Numbers
Section 2: *RL* (Parallel) Circuits in AC

Name: _____ Date: _____

NOTE ▶ Prior to performing this project and the upcoming projects that will use inductors, be sure to read the **Special Notes to Students and Instructors** at the beginning of Part 9. For this project, you will use the higher-frequency circuit option; therefore, pay attention to the cautions regarding DMM measuring limitations (frequency limits). If you are using a DMM, be sure it is rated to properly read voltages at the frequencies required in this project. If a scope is to be used, observe the critical ground connection precautions necessary to prevent shorting out components.

Procedure

1. Connect the circuit as shown.

2. Set the frequency of the function generator to 2,000 Hz and its output to 3 volts.

3. Measure V_A, V_{R_1}, V_{R_2}, V_{R_3}, and V_L and record the values in the Data Table, as appropriate.

4. From the measured data, calculate I_T (V_{R_1}/R_1), I_2 (V_{R_2}/R_2), I_L (V_{R_3}/R_3), and X_L (V_L/I_L) and record results in the Data Table in the appropriate locations.

5. Use the Pythagorean theorem formula and the measured and calculated values of I_2 (current through the resistive branch) and I_L (current through the inductive branch) to determine circuit I_T. Record values in the Data Table where indicated.

6. Use a dual-trace oscilloscope and perform a phase comparison of the current through the resistive branch (R_2) by observing V_{R_2} and the current through the inductor branch by observing V_{R_3}.

! CAUTION: Be sure the signal source ground and the scope ground(s) are connected to the same end (the bottom end) of the circuit network when making the measurements to prevent the grounds from shorting out a portion of the circuit!

Determine the phase difference between the two signals. If possible, demonstrate your scope waveforms to your instructor.

7. After completing the Data Table, answer the Analysis Questions and develop a brief Technical Lab Report to complete the project.

Data Table

Component and Parameter I.D.'s	Measured Values	Calculated Values
V_A (V)		—
V_{R_1} (V)		—
V_{R_2} (V)		—
V_{R_3} (V)		—
V_L (V)		—
I_T (mA)	—	
I_{R_2} (mA)	—	
X_L (kΩ)	—	
I_T, Pythagorean theorem (mA)	—	

Analysis Questions

NOTE Answers to these Analysis Questions should be clearly numbered and documented on separate sheets of paper with your name and the date at the top of each page. These answer sheets are to be turned in with the rest of the project documentation, as appropriate.

1. Explain why the total circuit current does not equal the arithmetic sum of the branch currents in this circuit.

2. Are the current through R_2 and the voltage across R_2 in phase with each other?

3. Define the phase relationship between the current through the inductor and the voltage across the inductor.

4. Using the data from your Data Table, draw an appropriate V-I vector diagram showing the branch currents and total current in the circuit. Label all vectors appropriately.

5. Use trigonometry to determine the circuit phase angle. Add the angle data to your vector diagram.

6. Explain whether the circuit total current is leading or lagging the circuit applied voltage and by how much. Also, explain why there is a phase difference in this circuit.

7. Show your work for determining the difference in phase between the resistive and inductive branch currents, based on your scope patterns. Was the phase difference close to 90 degrees? What might cause it to not be precisely 90 degrees?

8. Define and list the changes that would occur (in general terms) for the various circuit parameters if the source frequency were doubled, while keeping the V_A at 3 volts. (Define each specific parameter in terms of whether it would increase, decrease, or remain the same.)

Technical Lab Report

Write a brief technical lab report summarizing the technical facts learned from this project. The report should be organized to provide the following:

1. An introductory paragraph describing the type of circuit being analyzed and the key parameters that will be discussed relating to this circuit.

2. A section describing the most important characteristics of this type of circuit that were shown via the collected data in the tables and graphs.

3. Any special facts or characteristics about this type of circuit that were highlighted in answering the Analysis Questions.

4. A practical example of how the information learned in this project might help you in operating, troubleshooting, error analysis, or adjusting a circuit of this type in your home setting, in your training program setting, or in a job setting in the real world.

5. A summary statement listing the most positive aspects of the project and any parts of the project that were difficult because of equipment problems or unclear instructions. Include areas that might be improved.

PART 11

Summary
RL Circuits in AC

Name: _____ Date: _____

Complete the review questions, indicating the appropriate response by placing a check in the box next to the correct answer.

1. An ac circuit whose phase angle is 45 degrees is composed of
   - ☐ an equal amount of resistance and reactance
   - ☐ an unequal amount of resistance and reactance
   - ☐ pure resistance
   - ☐ pure reactance

2. The current through an inductor
   - ☐ leads V_L
   - ☐ lags V_L
   - ☐ is in phase with V_L

3. To find the Z of a series RL circuit
   - ☐ simply add X_L total and R total
   - ☐ subtract X_L from R
   - ☐ neither of these

4. When making a V-I vector diagram of a series RL circuit, the reference vector is
   - ☐ I_T
   - ☐ V_T
   - ☐ Z_T
   - ☐ R_T
   - ☐ none of these

5. When making a V-I vector diagram of a parallel RL circuit, the reference vector is
   - ☐ I_T
   - ☐ V_T
   - ☐ Z_T
   - ☐ R_T
   - ☐ none of these

6. In a series RL circuit, if L is increased while R remains the same, the circuit phase angle (angle between V_A and I_T) will
   - ☐ increase
   - ☐ decrease
   - ☐ remain the same

7. In a series RL circuit, if R is increased while L remains the same, the circuit phase angle will
   - ☐ increase
   - ☐ decrease
   - ☐ remain the same

8. In a parallel RL circuit, if L is increased while R remains the same, the circuit phase angle will
   - ☐ increase
   - ☐ decrease
   - ☐ remain the same

9. In a parallel *RL* circuit, if *R* is increased while *L* remains the same, the circuit phase angle will
   - ☐ increase
   - ☐ remain the same
   - ☐ decrease

10. The value of total circuit impedance (*Z*) for both series and parallel ac circuits can be solved by Ohm's law.
    - ☐ True
    - ☐ False

BASIC TRANSFORMER CHARACTERISTICS

Objectives

In these projects you will connect circuits that will illustrate the concepts of step-up and step-down transformer actions.

In completing these projects, you will connect circuits, make measurements, perform calculations, draw conclusions, and be able to answer questions about the following items relative to transformers:

- Transformer turns ratios
- Transformer voltage ratios
- Transformer current ratios
- The relationship of transformer impedances to turns and/or voltage ratios

Project/Topic Correlation Information

PROJECT		TEXT CHAPTER	SECTION	RELATED TEXT TOPIC(S)
45	**Turns, Voltage, and Current Ratios**	16	16-5	Important Transformer Ratios
46	**Turns Ratios versus Impedance Ratios**	16	16-5	Impedance Ratio

Basic Transformer Characteristics
Turns, Voltage, and Current Ratios

PROJECT 45

Name: _____   Date: _____

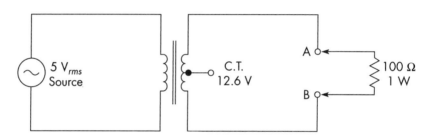

FIGURE 45-1 12.6-V center-tapped step-down transformer

FIGURE 45-2 Loaded transformer

PROJECT PURPOSE To demonstrate the relationships of turns ratios, voltage ratios, and current ratios in transformers by circuit measurements and observations.

PARTS NEEDED
- ☐ DMM
- ☐ Low-voltage source
- ☐ 12.6-V C.T. step-down transformer
- ☐ CIS
- ☐ Resistors 100 Ω, 1 W

SPECIAL NOTE:

Transformers may be used to "step up" or "step down" voltages. If it were possible to have 100% efficiency, all of the power in the primary would be transferred to the secondary. If this were true, then the product of V and I in the secondary would equal the product of V and I in the primary, and the current step-up or step-down ratio would be inverse to the voltage step-up or step-down ratio in order for the $V \times I$ products to be the same.

To illustrate the concepts of basic transformer action, we want you to assume 100% efficiency of the transformers used for any calculations you are required to perform.

! Caution: When working with HIGH VOLTAGE, use all the safety procedures relative to working with high voltage.

PROCEDURE

1. Obtain a 12.6-volt transformer and connect the circuit shown in Figure 45-1.

2. Use a voltmeter and measure the voltage present on the primary and on the secondary.

   ⚠ OBSERVATION   Primary V = _____ V.

   Secondary V = _____ V.

   ⚠ CONCLUSION   The primary-to-secondary voltage ratio is _____.

   The secondary-to-primary voltage ratio is _____.

   What is the N_S/N_P turns ratio? _____. Are the voltage and turns ratios the same? _____.

3. Connect a 100-Ω, 1-W resistor between points A and B as shown in Figure 45-2. Use Ohm's law to determine the secondary current.

   ⚠ OBSERVATION V_R = _____ V.

   ⚠ CONCLUSION $I = V/R$

   I = _____ A.

   Is the secondary current the same as the current through the load R? _____.

4. Use the previously determined voltage ratio from above and calculate the primary-to-secondary current ratio.

   ⚠ OBSERVATION $V_P:V_S$ = _____.

   $I_P:I_S$ = _____.

   ⚠ CONCLUSION   The current ratio is the _____ of the voltage ratio.
   Using the secondary current previously determined in Figure 45-2, step 3, and the current ratio just calculated, what is the approximate primary current value?
   _____ A.

Basic Transformer Characteristics
Turns Ratios versus Impedance Ratios

PROJECT 46

Name: _____ Date: _____

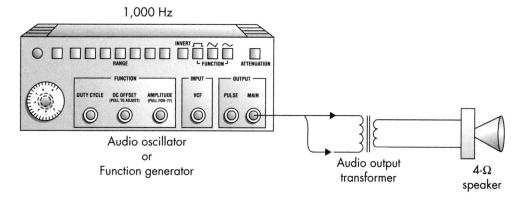

FIGURE 46-1

FIGURE 46-2 Optional circuit

PROJECT PURPOSE To demonstrate that the impedance ratio of a transformer is related to the square of the turns ratio.

PARTS NEEDED
- ☐ DMM
- ☐ Function generator or audio oscillator
- ☐ CIS
- ☐ Audio output transformer
- ☐ 4-Ω speaker

SPECIAL NOTE:

The impedance ratio of a transformer is related to the square of the turns ratio. This project will let you determine the turns ratio of a transformer, then theoretically determine the impedance ratio. Once this has been determined, you or your instructor will compare your results with the catalog specifications for the transformer you used. You can then see how closely your measurements and calculations correlate with the transformer specifications.

233

PROCEDURE

1. Obtain an audio output transformer and connect a circuit similar to that shown in Figure 46-1.

2. Apply a 1,000-Hz signal at a 3–5 V voltage level to the transformer primary. Measure the secondary voltage with an appropriate measuring instrument.

   ⚠ OBSERVATION   Voltage applied to the primary is _____ V.

   Measured voltage on secondary is _____ V.

   ⚠ CONCLUSION   The N_P/N_S ratio must be about _____ :1.

   Using the formula $Z_P/Z_S = N_P^2/N_S^2$, the primary-to-secondary impedance ratio is _____ :1.

3. Assume that the secondary load is going to be 4 ohms. Using the impedance ratio previously calculated, compute the nominal impedance of the circuit that should be connected to the primary of the output transformer.

   ⚠ OBSERVATION   Nominal Z calculated = _____ Ω.

   ⚠ CONCLUSION   Using the catalog parameters for the transformer, the primary-to-secondary Z ratio is _____ :1.

   Does the impedance ratio you measured and calculated via voltage measurements reasonably agree with the catalog specifications for the transformer? _____.
   What might cause any differences? _____ .

Optional Step

4. Connect a 4-Ω speaker to the secondary, Figure 46-2. Measure the voltages and determine the Z ratio.

   ⚠ OBSERVATION   Primary V = _____ V.   Calculated turns ratio = _____.
   Secondary V = _____ V.   Calculated Z ratio = _____.

   ⚠ CONCLUSION   Does this measured and calculated Z ratio more closely match the transformer specifications? _____ .

Story Behind the Numbers
Basic Transformer Characteristics

Name: _____ Date: _____

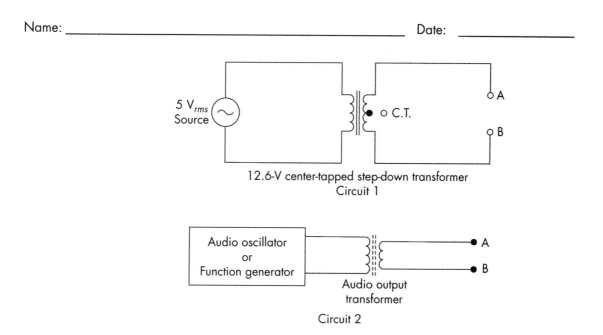

12.6-V center-tapped step-down transformer
Circuit 1

Circuit 2

Procedure

1. Obtain a 12.6-V transformer and connect Circuit 1 as shown.

2. Use a DMM and measure the voltage present at the primary and at the secondary (with no load connected to the secondary). Record readings in the Data Table, as appropriate.

3. Calculate the primary-to-secondary *voltage ratio* and log results in the Data Table.

4. Determine the approximate primary-to-secondary *turns ratio* and log results in the Data Table.

5. Connect a 100-Ω, 1-W resistor as a load across the secondary at points A and B.

6. Measure the voltage across the 100-Ω resistor and, by using Ohm's law, calculate the secondary current with the load.

7. Using the previously determined voltage ratio, determine the primary-to-secondary *current ratio,* and log results in the Data Table.

8. From your previous steps, determine the approximate primary current.

9. Using normal precautionary measures, temporarily remove voltage applied to the primary. Insert one DMM in the *current measuring function* and measure the primary current. Insert a second DMM to measure the

secondary current with the 100-Ω load. Record these measured currents in the Data Table, as appropriate.

10. Connect Circuit 2 as shown.

11. Apply a 3-V 1,000-Hz signal to the primary. Measure the secondary voltage (unloaded) and record the value in the Data Table, as indicated.

12. From your measurements, calculate and fill in the values for the primary-to-secondary turns ratio. From that data, determine the primary-to-secondary impedance ratio (Z_P/Z_S).

13. After completing the Data Table, answer the Analysis Questions and develop a brief Technical Lab Report to complete the project.

Data Table

Component and Parameter I.D.'s	Measured Values	Calculated Values
Circuit 1, primary V		—
Circuit 1, secondary V		—
P-S voltage ratio	—	
V_{R_L} @ points A–B		—
I secondary w/100Ω	—	
P-S current ratio	—	
Approx. I primary current	—	
I measured, primary		—
I measured, secondary		—
Circuit 2, secondary V		—
P-S turns ratio	—	
P-S Z ratio, calculated	—	

Analysis Questions

NOTE Answers to these Analysis Questions should be clearly numbered and documented on separate sheets of paper with your name and the date at the top of each page. These answer sheets are to be turned in with the rest of the project documentation, as appropriate.

1. Describe how the voltage ratio and current ratio of a transformer are related.

2. For this project, are we using the transformer in Circuit 1 as a step-up or a step-down transformer? Describe how this transformer might be used in the opposite way.

3. What percentage of transformer efficiency was assumed in determining the various transformer ratios from your voltage measurements?

4. By observing the parameters spelled out for transformers in an electronics parts catalog, list the electrical parameters you think are important when considering replacement of a transformer in an electronic circuit that is similar to the transformer used in project Circuit 1.

5. By observing the parameters spelled out for transformers in an electronics parts catalog, list the electrical parameters you think are important when considering replacement of a transformer in an electronic circuit that is similar to the transformer used in project Circuit 2.

6. For the transformer used in Circuit 2, assume that the secondary load is going to be 4 ohms. Using the impedance ratio previously calculated, compute the nominal impedance of the circuit that should be connected to the primary of the audio transformer.

Technical Lab Report

Write a brief technical lab report summarizing the technical facts learned from this project. The report should be organized to provide the following:

1. An introductory paragraph describing the type of circuit being analyzed and the key parameters that will be discussed relating to this circuit.

2. A section describing the most important characteristics of this type of circuit that were shown via the collected data in the tables and graphs.

3. Any special facts or characteristics about this type of circuit that were highlighted in answering the Analysis Questions.

4. A practical example of how the information learned in this project might help you in operating, troubleshooting, error analysis, or adjusting a circuit of this type in your home setting, in your training program setting, or in a job setting in the real world.

5. A summary statement listing the most positive aspects of the project and any parts of the project that were difficult because of equipment problems or unclear instructions. Include areas that might be improved.

Summary
Basic Transformer Characteristics

Name: _____ Date: _____

Answer the following questions with "T" for true and "F" for false. (Put your answer in the appropriate blank.)

1. A transformer's voltage and current ratios are inverse. _____

2. The impedance ratio of a transformer is related to the square root of the turns ratio. _____

3. If a transformer has a 3:1 turns ratio (secondary-to-primary) and a primary impedance of 5,000 ohms, the load impedance on the secondary should be 45,000 ohms to affect a proper impedance match. _____

4. If a transformer's turns ratio is doubled, the related impedance ratio will quadruple. _____

5. A transformer with a secondary-to-primary turns ratio of 10:1 will have a secondary voltage of 50 volts with 0.5 volts applied to its primary. _____

6. If a transformer's current ratio is 1:3, then the related voltage ratio will be 1:3. _____

7. If a transformer's voltage ratio is doubled, its related current ratio will halve. _____

8. If the turns on a transformer secondary are doubled, the impedance of the secondary will be four times as great as the original impedance. _____

9. Transformer efficiency is typically 100%. _____

10. The voltage across a transformer secondary will be higher when it is "unloaded" as compared to when it is supplying current to a load. _____

CAPACITANCE (DC CHARACTERISTICS)

PART 13

Objectives

You will connect circuits illustrating the characteristics of capacitor(s) in dc circuits.

In completing these projects, you will connect circuits, make measurements, perform calculations, draw conclusions, and be able to answer questions about the following items related to capacitance in dc circuits:

- Charge and discharge action
- *RC* time
- Total capacitance of series capacitors
- Total capacitance of parallel capacitors

Project/Topic Correlation Information

PROJECT	TEXT CHAPTER	SECTION	RELATED TEXT TOPIC(S)
47 Charge and Discharge Action and *RC* Time	17	17-3 17-10	Charging and Discharging Action The *RC* Time Constant
48 Total Capacitance in Series and Parallel	17	17-8	Total Capacitance in Series and Parallel

Capacitance (DC Characteristics)
Charge and Discharge Action and *RC* Time

PROJECT 47

Name: _____   Date: _____

FIGURE 47-1

PROJECT PURPOSE To demonstrate the charging and discharging action of a capacitor and to observe that a capacitor takes five *RC* time constants to change from one set voltage level to another.

PARTS NEEDED
- ☐ DMM
- ☐ VVPS (dc)
- ☐ CIS
- ☐ Capacitors
  0.1 µF, 1.0 µF
- ☐ Resistors
  100 kΩ
  1 MΩ
  10 MΩ

SAFETY HINTS Remember that charged capacitors remain charged and can shock you (until they are discharged).

PROCEDURE

1. Connect the initial circuit as shown in Figure 47-1.

2. Set V_A at 20 volts. Insert a jumper wire between points A and B and observe the meter action.

   ⚠ **OBSERVATION** Current flowed for approximately _____ seconds as evidenced by the voltage measured across R. The rate of charge was (*linear, nonlinear*) _____.

   ⚠ **CONCLUSION** The current that flowed was the charging current that charged the capacitor to a voltage equal to (V_R, V_A) _____. Was charging current maximum at the beginning of the charge time or near the end? _____. At the beginning of the charge time, the voltage across R was equal to (V_C, V_A) _____ (the first instant). At the end of the charge time, the voltage across the resistor is _____ volts; the voltage across the capacitor is equal to V_A.

243

NOTE ▶ The charged capacitor voltage is equal to V_A and series opposing the source. Hence, no current can flow once the capacitor is charged.

3. Remove the jumper from points A and B. Reverse the polarity of the voltmeter. Insert a jumper between points C and D and observe the meter action during discharge of the capacitor.

    ⚠ **OBSERVATION**   Discharge time was approximately _____ seconds. The *rate* of discharge was (*linear, nonlinear*) _____.

    ⚠ **CONCLUSION**   Did the capacitor take the same time to discharge through R as it did to charge? _____. At the end of the discharge time V_C = _____ ; V_R = _____ V.

4. Remove the jumper from points C and D. Change the DMM polarity from the original setup if it is not an "autopolarity"-type meter. Change R to a 10-MΩ resistor. Repeat the sequence of steps 2 and 3.

    ⚠ **OBSERVATION**   Charge time was approximately _____ seconds. Discharge time was _____ seconds.

    ⚠ **CONCLUSION**   Increasing R increased the charge time because the charging current was limited to a smaller value. Thus, it took longer to obtain a given potential difference or _____ across the C.

5. Change the C value to a 0.1-μF capacitor. Insert a jumper between points A and B and note the charge time.

    ⚠ **OBSERVATION**   Charge time was approximately _____ seconds.

    ⚠ **CONCLUSION**   Changing C from a 1.0-μF to a 0.1-μF capacitor caused the charge time to _____. We conclude that both the value of _____ and of _____ determine charge and discharge time. It should be noted that if it were not for the effect of the multimeter resistance, the time to charge the capacitor or discharge it would be directly proportional to R and to C. The formula relating to this is called the formula for the *time constant*. This formula states that one RC time constant = R in ohms times C in farads, and the answer is in seconds.

    Also, it should be observed that it takes five time constants to charge or discharge the capacitor. Calculate the R_e of R and the meter circuit's resistance in parallel and determine if the charge time is about equal to the expected 5 TC. Is it? _____.

6. Use the RC time constant formula and determine how long it would take to charge a 1-μF capacitor in series with a 100-kΩ resistor.

    ⚠ **OBSERVATION**   5 RC time constants = _____ seconds.

7. Connect the circuit described in step 6 and note the charge time.

    ⚠ **OBSERVATION**   Charge time measured approximately _____ seconds.

    ⚠ **CONCLUSION**   A multimeter with a meter circuit R of 5 MΩ or greater does not alter the circuit resistance of 100 kΩ very much. The measured charge time was (*close, not close*) _____ to the theoretical value.

Capacitance (DC Characteristics)
Total Capacitance in Series and Parallel

PROJECT 48

Name: _____ Date: _____

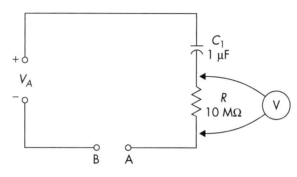

FIGURE 48-1

PROJECT PURPOSE To demonstrate that capacitors in series add like resistances in parallel and that capacitors in parallel add like resistances in series, using circuit observations or *RC* times.

PARTS NEEDED
- ☐ DMM
- ☐ VVPS (dc)
- ☐ CIS
- ☐ Capacitor 1.0 µF (2)
- ☐ Resistor 10 MΩ

 SAFETY HINTS REMEMBER: Capacitors hold their charge unless a discharge path is provided! Don't let that discharge path be you!

SPECIAL NOTE:

For this project, we will take advantage of the fact that the charge time of a capacitor is directly proportional to capacitance. By noting the charge time of a single capacitor, then noting the charge time for two capacitors in series and then in parallel, we should be able to conclude the effect on total capacitance of connecting capacitors in series or parallel. A DMM is highly preferred over a VOM for voltage measurements here.

PROCEDURE

1. Connect the initial circuit as shown in Figure 48-1.

2. Set V_A to 20 volts. Insert a jumper between points A and B and note the charge time by observing the voltmeter.

⚠ **OBSERVATION**   Charge time was approximately _____ seconds.

⚠ **CONCLUSION**   One *RC* time is approximately _____ seconds. This means that the R_e of the meter and the 10-MΩ resistor is approximately _____ Ω.

3. Obtain a second 1-µF capacitor. Remove the jumper from points A and B. Carefully discharge C_1, then insert the second capacitor (C_2) in series with C_1. Insert a jumper between points A and B and note the charge time of C_1 and C_2 in series.

   ⚠ **OBSERVATION**   Charge time was approximately _____ seconds.

   ⚠ **CONCLUSION**   Since the charge time has decreased to _____ the value it was with only C_1 in the circuit, it may be concluded that the total capacitance has (*increased, decreased*) _____. Since the new *RC* time is (*double, half*) _____ the original and the *R* has not been changed, we conclude that the new total capacitance is _____ µF. Our observations tell us that capacitors in series add like resistors in (*series, parallel*) _____.

4. Remove the jumper from points A and B. Discharge the capacitors. Change the circuit as required to achieve a circuit with C_1 and C_2 in parallel, and this combination in series with *R*. Insert the jumper again, and note the charge time of C_1 and C_2 in parallel.

   ⚠ **OBSERVATION**   Charge time was approximately _____ seconds.

   ⚠ **CONCLUSION**   The charge time for this step is approximately (*2, 3, 4*) _____ times the time recorded in step 2. This indicates that the total capacitance of C_1 and C_2 in parallel is (*1/2, 2 ×, 3 ×*) _____ the capacitance of C_1 alone. We may conclude that capacitors in parallel add like resistors in (*series, parallel*) _____. If C_2 were not the same value as C_1, would the statements concerning total capacitance of capacitors in series and parallel still hold true? (That is, series *C*s add like parallel *R*s, and parallel *C*s add like series *R*s.) _____.

Story Behind the Numbers
Capacitance (DC Characteristics)

Name: _____   Date: _____

Procedure

1. Connect the circuit as shown.

2. Connect a DMM to measure the voltage across R. Set V_A at 20 volts. While observing and timing the meter action, carefully place a jumper lead between points A and B and time the approximate number of seconds it takes for the meter to reach a zero-volt reading. (This is the approximate capacitor "charge time.") Record this time in the Data Table.

3. Carefully remove the jumper lead from between points A and B. *If* your meter needs polarity protection, reverse the polarity of meter connections across R. Now insert a jumper lead between points C and D while observing the approximate time it takes to discharge the capacitor. Record this time in the Data Table.

4. Turn off the power supply and remove the jumper lead between points C and D.

5. Modify the circuit by changing the R value from 1 MΩ to 10 MΩ and the C value from 1 µF to 0.1 µF.

6. Again, connect a DMM to measure the voltage across R. Set V_A at 20 volts. While observing and timing the meter action, carefully place a jumper lead between points A and B and time the approximate number of seconds it takes for the meter to reach a zero-volt reading. (This is the approximate capacitor "charge time.") Record this time in the Data Table.

7. Turn off the power supply and remove the jumper lead between points A and B. Remove the meter from the circuit. Jumper points C and D in order to discharge the capacitor.

8. Use the *RC* time constant formula and calculate the charge time for a circuit consisting of a 1-µF capacitor and a 100-kΩ resistor. Record this number in the appropriate location in the Data Table.

247

Data Table

Component and Parameter I.D.s	Measured Values	Calculated Values
V_A (V)	20	—
Charge time (1 µF & 1 MΩ) sec.		—
Discharge time (1 µF & 1 MΩ) sec.		—
Charge time (0.1 µF & 10 MΩ) sec.		—
Discharge time (0.1 µF & 10 MΩ) sec.		—
RC time (1-µF & 100-kΩ calc.) sec.		

9. Replace the 0.1-µF capacitor with a 1-µF capacitor and the 10-MΩ resistor with a 100-kΩ resistor. Again, connect a DMM to measure the voltage across R. Set V_A at 20 volts. While observing and timing the meter action, carefully place a jumper lead between points A and B and time the approximate number of seconds it takes for the meter to reach a zero-volt reading. (This is the approximate capacitor "charge time" for this circuit.) Record this time in the Data Table.

10. Turn off the power supply and remove the jumper lead between points A and B. Remove the meter from the circuit jumper points C and D in order to discharge the capacitor.

11. After completing the Data Table, answer the Analysis Questions and develop a brief Technical Lab Report to complete the project.

Analysis Questions

NOTE Answers to these Analysis Questions should be clearly numbered and documented on separate sheets of paper with your name and the date at the top of each page. These answer sheets are to be turned in with the rest of the project documentation, as appropriate.

1. Would it have taken the same amount of time to charge or discharge the capacitors in this circuit if V_A had been 30 volts instead of 20 volts? Explain.

2. If the capacitor had been "precharged" to 20 volts and V_A were changed to 30 volts, would it take the same amount of time for the capacitor to charge from 20 volts up to 30 volts as it did to charge from zero volts to 20 volts? Explain.

3. Were the charge and discharge times approximately equal when the circuit conditions were changed from a 1-µF capacitor and a 1-MΩ resistor to a 0.1-µF capacitor and a 10-MΩ resistor? Explain.

4. If a VOM were used for measurements rather than a DMM, would the time results have been different? Explain.

5. If the measurements would have been affected, which of the circuits would the VOM have affected the most?

6. How many RC time constants does it take for a capacitor to charge or discharge from one voltage level to another?

Technical Lab Report

Write a brief technical lab report summarizing the technical facts learned from this project. The report should be organized to provide the following:

1. An introductory paragraph describing the type of circuit being analyzed and the key parameters that will be discussed relating to this circuit.

2. A section describing the most important characteristics of this type of circuit that were shown via the collected data in the tables and graphs.

3. Any special facts or characteristics about this type of circuit that were highlighted in answering the Analysis Questions.

4. A practical example of how the information learned in this project might help you in operating, troubleshooting, error analysis, or adjusting a circuit of this type in your home setting, in your training program setting, or in a job setting in the real world.

5. A summary statement listing the most positive aspects of the project and any parts of the project that were difficult because of equipment problems or unclear instructions. Include areas that might be improved.

Summary
Capacitance (DC Characteristics)

Name: _____ Date: _____

Complete the following review questions, indicating the appropriate response by placing a check in the box next to the correct answer.

1. A charged capacitor has a difference of potential between its plates due to
   - ☐ an excess of electrons on one plate and a deficiency of electrons on the other
   - ☐ an excess of electrons on both plates
   - ☐ neither of these

2. A capacitor in a given circuit will take the same amount of time to discharge as it does to charge.
   - ☐ True
   - ☐ False

3. If the value of R is doubled and the value of C is halved in a given RC circuit, the time it will take to charge the capacitor will
   - ☐ increase
   - ☐ decrease
   - ☐ remain the same

4. One RC time constant is equal to
   - ☐ $R + C$
   - ☐ R/C
   - ☐ $R - C$
   - ☐ $R \times C$
   - ☐ none of these

5. In order for a capacitor to fully charge or discharge, it takes
   - ☐ one time constant
   - ☐ two time constants
   - ☐ four time constants
   - ☐ five time constants

6. For a given RC circuit, increasing the value of V_A will cause the time needed for the capacitor to fully charge or discharge to
   - ☐ increase
   - ☐ decrease
   - ☐ remain the same

7. The total capacitance of a 0.05-μF capacitor and a 0.1-μF capacitor in parallel is
   - ☐ 0.05 μF
   - ☐ 0.1 μF
   - ☐ 0.06 μF
   - ☐ 0.15 μF
   - ☐ 0.033 μF
   - ☐ none of these

8. The total capacitance of a 0.05-μF capacitor and a 0.1-μF capacitor in series is
   - ☐ 0.05 μF
   - ☐ 0.1 μF
   - ☐ 0.06 μF
   - ☐ 0.15 μF
   - ☐ 0.033 μF
   - ☐ none of these

9. How long would it take two parallel 0.1-μF capacitors to charge through a 1-MΩ resistance?
   - ☐ 0.2 second
   - ☐ 1 second
   - ☐ 0.1 second
   - ☐ 5 seconds
   - ☐ none of these

10. How long would it take two series 0.1-μF capacitors to charge through a 1-MΩ resistance?
    - ☐ 1/4 second
    - ☐ 1/2 second
    - ☐ 3/4 second
    - ☐ 1 second
    - ☐ 2.5 seconds
    - ☐ none of these

CAPACITIVE REACTANCE IN AC

PART 14

Objectives

You will connect several ac circuits illustrating the characteristics of capacitance in ac and the relationship of capacitive reactance to frequency and capacitance.

In completing these projects, you will connect circuits, make measurements, perform calculations, draw conclusions, and be able to answer questions about the following items related to capacitance and capacitive reactance in ac circuits:

- Characteristic(s) of capacitance in ac circuits
- Relationship of X_C to capacitance value
- Relationship of X_C to frequency
- Total reactance of series capacitors
- Total reactance of parallel capacitors
- The X_C Formula

Project/Topic Correlation Information

PROJECT		TEXT CHAPTER	SECTION	RELATED TEXT TOPIC(S)
49	Capacitance Opposing a Change in Voltage	18	18-2	V and I Relationships in a Purely Capacitative AC Circuit
50	X_C Related to Capacitance and Frequency	18	18-4	Relationship of X_C to Capacitance Value
			18-5	Relationship of X_C to Frequency of AC
51	The X_C Formula	18	18-6	Methods to Calculate X_C

Capacitive Reactance in AC
Capacitance Opposing a Change in Voltage

Name: _____ Date: _____

FIGURE 49-1

PROJECT PURPOSE To demonstrate the delaying effect a capacitor has on a change in voltage by varying a dc source level and observing the time it takes for the capacitor voltage to track the change in source voltage.

PARTS NEEDED
- ☐ DMM
- ☐ VVPS (dc)
- ☐ CIS
- ☐ Capacitor 1.0 µF
- ☐ Resistor 10 MΩ

SAFETY HINTS Remember the charged capacitor warning you have seen before!

PROCEDURE

1. Connect the initial circuit as shown in Figure 49-1.

2. Temporarily remove one lead from the VVPS and then set its output voltage at 10 volts. Observe how much time it takes the capacitor to stop charging after the lead is reinserted into the VVPS by watching the voltmeter measuring V_C.

   ⚠ OBSERVATION Approximate time to reach a steady state for V_C was _____ seconds.

   ⚠ CONCLUSION V_C did not take 50 seconds to charge because the (*wire, meter*) _____ resistance formed a voltage divider with the 10-MΩ resistor.

3. Quickly turn the VVPS voltage control knob to a higher voltage setting, and note how much time it takes V_C to reach its new steady-state value.

   ! CAUTION: Do not set V_A higher than the multimeter voltage range setting.

255

OBSERVATION Approximate time to reach the new steady-state value was _____ seconds.

CONCLUSION It takes the voltage on the capacitor (*the same, a different*) _____ amount of time to change from some given value to a new value, as it does for V_C to change from zero to any steady-state value.

4. Quickly turn the VVPS voltage control knob to a lower voltage setting and observe if V_C changes instantaneously, or takes time.

OBSERVATION It (*did, didn't*) _____ take time for V_C to decrease to its new steady-state value.

CONCLUSION From our observations, we conclude that a capacitor seems to oppose a change in (*current, voltage*) _____ in a similar fashion to a coil or inductor opposing a change in (*current, voltage*) _____.

Capacitive Reactance in AC
X_C Related to Capacitance and Frequency

Name: _____ Date: _____

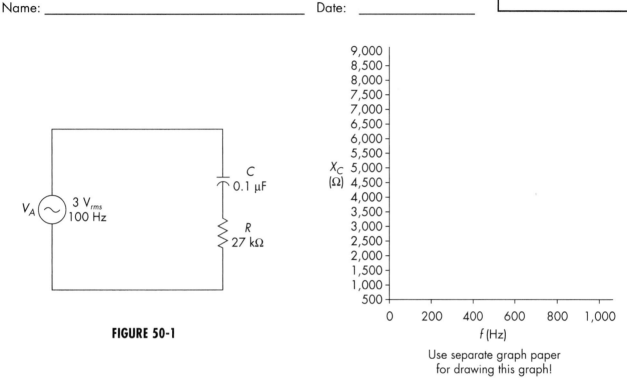

FIGURE 50-1

FIGURE 50-2 Sample graph coordinates

Use separate graph paper for drawing this graph!

PROJECT PURPOSE To verify the inverse relationship of X_C to both capacitance value and frequency by changing the values of circuit C and f, and measuring and analyzing the circuit parameter changes caused by changes in C and f.

PARTS NEEDED
- ☐ DMM
- ☐ Function generator or audio oscillator
- ☐ CIS
- ☐ Capacitors 0.1 µF   1.0 µF
- ☐ Resistor 27 kΩ

SAFETY HINTS Here's the charged capacitor warning, again! Just a reminder!

PROCEDURE

1. Connect the initial circuit as shown in Figure 50-1.

2. Set the function generator or audio oscillator to a frequency of 100 Hz and V_A to 3 volts. Measure V_R and V_C. Calculate the circuit I by Ohm's law, and also calculate X_C ($X_C = V_C/I$).

⚠ **OBSERVATION**

V_A = _____ V. I_T = _____ mA.

V_R = _____ V. X_C = _____ Ω.

V_C = _____ V.

257

PART 14: Capacitive Reactance in AC

⚠ CONCLUSION Since the current is the same through all parts of a series circuit and the voltage drop across R is approximately two times V_C, it appears that X_C must be about (*double, half*) _____ the value of R.

3. Change C to a 1.0-μF capacitor and repeat step 2.

   ⚠ OBSERVATION
   $V_A =$ _____ V. $I_T =$ approximately _____ mA.
   $V_R =$ _____ V. $X_C =$ approximately _____ Ω.
   $V_C =$ _____ V.

   ⚠ CONCLUSION Increasing the value of C while maintaining all other parameters the same caused X_C to (*increase, decrease*) _____. The X_C for the 1.0-μF capacitor was approximately (*ten times, one-tenth*) _____ the X_C of the 0.1-μF C. From this we conclude that X_C is (*directly, inversely*) _____ proportional to capacitance. This means as C decreases, X_C (*increases, decreases*) _____, or, if X_C has decreased, then C must have (*increased, decreased*) _____, all other factors being constant.

4. Keeping the 1.0-μF capacitor, change the frequency to 200 Hz and maintain 3 volts V_A. Repeat the measurements and calculations of the previous steps.

   ⚠ OBSERVATION
   $V_A =$ _____ V. $I_T =$ approximately _____ mA.
   $V_R =$ _____ V. $X_C =$ approximately _____ Ω.
   $V_C =$ _____ V.

   ⚠ CONCLUSION Increasing the frequency and keeping the C the same caused X_C to (*increase, decrease*) _____. If there were no voltmeter loading effects on the circuit, the indicated X_C at 200 Hz would have been (*two times, one-half*) _____ the X_C value at 100 Hz. From this we see that X_C is (*directly, inversely*) _____ proportional to frequency.

5. Keep f at 200 Hz and change C back to a 0.1-μF capacitor. Repeat the measurements and calculations of the previous steps.

   ⚠ OBSERVATION
   $V_A =$ _____ V. $I_T =$ _____ mA.
   $V_R =$ _____ V. $X_C =$ _____ Ω.
   $V_C =$ _____ V.

   ⚠ CONCLUSION Referring back to step 2 observations, does it appear that the X_C of the 0.1-μF C at 200 Hz (this step) is about one-half that at 100 Hz? _____. We conclude from the preceding that X_C is inversely proportional to _____ and _____.

6. Plot a graph of X_C versus f from 200 Hz to 1,000 Hz using the 0.1-µF capacitor.

   NOTE ➤ Calculate X_C for each 200-Hz change in frequency and plot with coordinates similar to those shown in Figure 50-2. Plot the graph on a separate sheet of graph paper.

   ⚠ OBSERVATION X_C at 200 Hz = _____. X_C at 800 Hz = _____.

   X_C at 400 Hz = _____. X_C at 1,000 Hz = _____.

   X_C at 600 Hz = _____.

   ⚠ CONCLUSION Did X_C act inversely proportional to f? _____.

Capacitive Reactance in AC
The X_C Formula

PROJECT 51

Name: _____   Date: _____

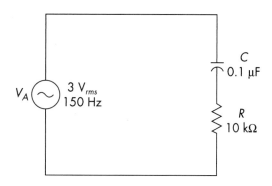

FIGURE 51-1

PROJECT PURPOSE To verify the X_C formula by changing C and f values in a simple RC circuit and making measurements and appropriate Ohm's law calculations to see if the formula is valid.

PARTS NEEDED
- ☐ DMM
- ☐ Function generator or audio oscillator
- ☐ CIS
- ☐ Capacitors 0.1 µF   1.0 µF
- ☐ Resistor 10 kΩ

SPECIAL NOTE:

As shown in the previous project, the X_C of a capacitor in ohms is inversely proportional to the frequency and also to the capacitance value. This relationship is shown by the X_C formula,

$$X_C = \frac{1}{2\pi f C}$$

The rationale we might use in understanding this formula is as follows. A larger C requires more charge (stored electrons) to arrive at a given difference of potential between its plates. This means that more charging (and discharging) current must flow in the circuit in order for the capacitor to "follow" the ac voltage applied to it at a given frequency than a smaller C would require under the same conditions. More current flowing for a given applied voltage indicates a lower opposition. A higher frequency also causes more charge and discharge current to flow per unit time in order for the voltage across the capacitor to follow the applied voltage.

PROCEDURE

1. Connect the initial circuit as shown in Figure 51-1.

2. Set the function generator at a frequency of 150 Hz and V_A at 3 volts. Measure V_R and then calculate I_T. Measure V_C and use the calculated I to determine X_C by Ohm's law. Calculate X_C by the X_C formula.

   ⚠ OBSERVATION
   $V_A =$ _____ V. $I =$ _____ mA.
   $V_R =$ _____ V. $V_C =$ _____ V.
   X_C by Ohm's law approximately = _____ Ω.
   X_C by X_C formula = _____ Ω.

   ⚠ CONCLUSION
   Since this is a series circuit and V_C is virtually the same as V_R, it is apparent that the X_C must essentially be equal to _____. Was the X_C calculated by the capacitative reactance formula close to the value determined by Ohm's law? _____.

3. Change the frequency to 75 Hz and keep V_A at 3 volts. Make the measurements and calculations described in step 2 above.

   ⚠ OBSERVATION
   $V_A =$ _____ V. $I =$ _____ mA.
   $V_R =$ _____ V. $V_C =$ _____ V.
   X_C by Ohm's law approximately = _____ Ω.
   X_C by X_C formula = _____ Ω.

   ⚠ CONCLUSION
   Lowering the frequency to one-half its previous value caused the X_C to (*increase, decrease*) _____ to a value nearly (*double, half*) _____ the original. Two reasons for the X_C values calculated by Ohm's law method and X_C formula not being the same might be the resistor and capacitor _____ and meter _____ effects.

4. Keep the frequency at 75 Hz, but change C to 1 µF. Measure and calculate as before.

   ⚠ OBSERVATION
   $V_A =$ _____ V. $I =$ _____ mA.
   $V_R =$ _____ V. $V_C =$ _____ V.
   X_C by Ohm's law approximately = _____ Ω.
   X_C by X_C formula = _____ Ω.

   ⚠ CONCLUSION
   Increasing C by a factor of ten caused the X_C to (*increase, decrease*) _____. Did the X_C approximately change by a factor of ten? _____. From the observations we have made, does it appear that the X_C formula is functional for predicting parameters in practical circuits? _____.

Story Behind the Numbers
Capacitive Reactance in AC

Name: _____ Date: _____

Initial Circuit

Source: 3 V$_{rms}$, 100 Hz
C = 0.1 μF
R = 27 kΩ

Procedure

NOTE When performing the procedure steps, you have the option of using our data collection tables and creating any requested graphs by hand or you may use Excel for creating the required tables and create any requested graphs from the data in your Excel tables using the Excel chart feature.

1. Connect the initial circuit as shown.

2. Set the function generator (source) to a frequency of 100 Hz and V_A to 3 volts. Measure V_R and V_C. Calculate the circuit current and the value of X_C using your measurements and Ohm's law, as appropriate. Record measured and calculated data in the Data Table, as appropriate. Disconnect one end of the circuit from the source.

3. Change the capacitor to a 1-μF capacitor. Reconnect the circuit to the source, again measure the values of V_R and V_C, and calculate circuit current and the value of X_C. Record data in the Data Table, as indicated.

4. Change the C value back to 0.1 μF, using appropriate precautions. Keep the signal frequency at 200 Hz and the input voltage set at 3 volts. Make a "frequency run" that will illustrate X_C versus f. Start at 200 Hz and incrementally change frequency by 200-Hz steps up to 1,000 Hz. (Check that V_A is still at 3 volts at each new frequency setting.) Calculate X_C for each new frequency setting, as you did before. (That is, determine circuit I value from V_R/R, then divide V_C by the I value to determine X_C.) Log all data in the Data Table, as appropriate.

5. Use the data from your frequency run and create a line graph of X_C versus frequency (f).

Data Table

───── Frequency Run ─────▶

Component and Parameter I.D.s	0.1 µF @ 100 Hz	1.0 µF @ 100 Hz	f_1 0.1 µF @ 200 Hz	f_2 0.1 µF @ 400 Hz	f_3 0.1 µF @ 600 Hz	f_4 0.1 µF @ 800 Hz	f_5 0.1 µF @ 1,000 Hz
V_A (V)	3	3	3	3	3	3	3
V_R measured (V_{rms})							
V_C measured (V_{rms})							
I_T (ac) calculated (mA)							
X_C calculated (kΩ)							

6. After completing the Data Table and creating the line graph, answer the Analysis Questions and develop a brief Technical Lab Report to complete the project.

Analysis Questions

1. From observing your data, define the relationship of capacitance value to capacitive reactance at any given frequency. Show a specific example of your conclusion.

2. From observing your data, define the relationship of capacitive reactance to frequency for any given capacitance value. Show at least one specific example of your conclusion.

3. From observing your graph of X_C versus frequency, does X_C change value linearly or nonlinearly with frequency change? Explain your conclusion.

4. From observing your graph of X_C versus frequency, would you say that capacitive reactance has a direct or an inverse relationship to frequency? Show a specific example of your conclusion.

5. Write the formula for finding X_C that shows these direct or inverse relationships between C, f, and capacitive reactance.

6. Show the form of the X_C formula that will allow you to find the capacitance value if you know the frequency and the capacitive reactance values.

Technical Lab Report

Write a brief technical lab report summarizing the technical facts learned from this project. The report should be organized to provide the following:

1. An introductory paragraph describing the type of circuit being analyzed and the key parameters that will be discussed relating to this circuit.

2. A section describing the most important characteristics of this type of circuit that were shown via the collected data in the table and graphs.

3. Any special facts or characteristics about this type of circuit that were highlighted in answering the Analysis Questions.

4. A practical example of how the information learned in this project might help you in operating, troubleshooting, error analysis, or adjusting a circuit of this type in your home setting, in your training program setting, or in a job setting in the real world.

5. A summary statement listing the most positive aspects of the project and any parts of the project that were difficult because of equipment problems or unclear instructions. Include areas that might be improved.

Summary
Capacitive Reactance in AC

Name: _____   Date: _____

Complete the following review questions, indicating the appropriate response by placing a check in the box next to the correct answer.

1. X_C is measured in ohms because it limits ac current to a value of
   - ☐ $0.159 \times f \times C$
   - ☐ V/X_C
   - ☐ $0.159/fC$
   - ☐ none of these

2. The formula for X_C is
   - ☐ $X_C = 2\pi/fC$
   - ☐ $X_C = 1/2\pi fC$
   - ☐ $X_C = fC/0.159$
   - ☐ none of these

3. For a given value of C, if f is increased, then X_C will
   - ☐ increase
   - ☐ decrease
   - ☐ remain the same

4. For a given value of f, if C is decreased, then X_C will
   - ☐ increase
   - ☐ decrease
   - ☐ remain the same

5. A capacitor appears to oppose a change in voltage because current must flow before a difference of potential can be established across a capacitor.
   - ☐ True
   - ☐ False

6. Since capacitors in series add like resistors in parallel, the X_C of two series capacitors will be
   - ☐ greater than either one alone
   - ☐ less than either one alone

7. Capacitive reactances in parallel add like resistances in
   - ☐ series
   - ☐ parallel
   - ☐ neither of these

8. If C and f are both doubled in a given circuit, the X_C will
   - ☐ increase two times
   - ☐ decrease two times
   - ☐ increase four times
   - ☐ decrease four times

9. If C is doubled and f is halved in a given circuit, the X_C will
   - ☐ increase
   - ☐ decrease
   - ☐ remain the same

10. What is the total capacitive reactance of two series 1-μF capacitors at a frequency of 200 Hz?
    - ☐ 3,180 Ω
    - ☐ 15.9 Ω
    - ☐ 318 Ω
    - ☐ 1.59 kΩ
    - ☐ 31.8 kΩ
    - ☐ none of these

RC CIRCUITS IN AC

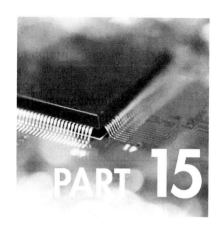

Objectives

You will connect several ac *RC* circuits and make measurements and observations regarding their important electrical characteristics.

In completing these projects, you will connect circuits, make measurements, perform calculations, draw conclusions, and be able to answer questions about the following items related to *RC* circuits:

- Relationship of circuit phase angle to R and X_C
- Relationships of current and voltage for capacitors and resistors
- Circuit impedance
- Simple vector diagram(s)
- Reference vector(s)

Project/Topic Correlation Information

PROJECT	TEXT CHAPTER	SECTION	RELATED TEXT TOPIC(S)
52 *V*, *I*, *R*, *Z*, and θ Relationships in a Series *RC* Circuit	19	19-2	Series *RC* Circuit Analysis
53 *V*, *I*, *R*, *Z*, and θ Relationships in a Parallel *RC* Circuit	19	19-4	Parallel *RC* Circuit Analysis

RC Circuits in AC
V, I, R, Z, and θ Relationships in a Series RC Circuit

PROJECT 52

Name: _____ Date: _____

FIGURE 52-1

(Circuit: V_A = 3 V_{rms}, 100 Hz; C = 0.1 μF; R = 27 kΩ, series)

PROJECT PURPOSE To demonstrate the key electrical parameter relationships in a series *RC* circuit. To observe that due to out-of-phase elements, simple dc analysis techniques cannot be used to determine circuit parameters in ac circuits containing reactive components. Furthermore, to provide practice in using simple ac analysis techniques and in drawing ac circuit vector diagrams.

PARTS NEEDED
- ☐ DMM
- ☐ Function generator or audio oscillator
- ☐ Dual-trace oscilloscope
- ☐ CIS
- ☐ Capacitor 0.1 μF
- ☐ Resistor 27 kΩ

SPECIAL NOTE:

By way of review, recall that in a purely resistive series ac circuit, the circuit current was in phase with the applied voltage. Also, the voltage drops across the individual resistors were in phase with current and with each other. The total opposition to current flow in the purely resistive circuit of this type was the arithmetic sum of the individual resistances. If a vector diagram of voltage and current were drawn, we would use the current as the reference vector, since it is the common factor in a series circuit. Summarizing: $Z = R_T$, $\theta = 0$ degrees, V_T = simple sum of $V_1 + V_2 + ...$ (etc.) for purely resistive series ac circuits. In this project, you will examine parameters for a series *RC* circuit.

PROCEDURE

1. Connect the initial circuit as shown in Figure 52-1.

2. Set the frequency of the function generator to 100 Hz and V_A to 3 volts. Measure V_A, V_R, and V_C.

   ⚠ OBSERVATION $V_A = $ _____ V. $V_C = $ _____ V.

   $V_R = $ _____ V.

   ⚠ CONCLUSION   Does the sum of V_R and V_C equal V_A? _____. This is because V_R and V_C are (*in phase, out of phase*) _____.

3. Calculate I_T from V_R/R; X_C from V_C/I, and Z from V_T/I_T.

   ⚠ OBSERVATION $I_T = $ _____ mA. $Z = $ _____ Ω.

   $X_C = $ _____ Ω.

   ⚠ CONCLUSION   Does Z equal the arithmetic sum of R and X_C? _____. We may conclude that since V_A is not equal to $V_R + V_C$ and Z is not equal to $R + X_C$, these values are the resultant of two out-of-phase vectors. We can solve for the resultant vectors by means of the Pythagorean theorem or trigonometry. Is the voltage across the resistor in phase with the current through it? _____. Since a capacitor opposes a change in voltage, we might assume that the I_C (*leads, lags*) _____ V_C. In a perfect capacitor I_C (*leads, lags*) _____ V_C by 90 degrees. We should also conclude that since this is a series circuit, the larger the amount of X_C (the smaller the C), the more like a pure (*resistive, capacitive*) _____ circuit the circuit will act. This means the larger the X_C compared to the R, the (*greater, smaller*) _____ will be the resultant phase angle between V_A and I_T.

4. Calculate Z using the Pythagorean approach.

   ⚠ OBSERVATION Z calculated = _____ Ω.

5. Draw an impedance diagram for the circuit in the "Observation" section.

   ⚠ OBSERVATION

6. Use the Z diagram and trigonometry to determine the phase angle.

   OBSERVATION   θ = _____ degrees.

7. Repeat the previous steps 2 through 6. This time set $f = 200$ Hz and keep V_A at 3 volts.

   OBSERVATION V_A = _____ V. X_C = _____ Ω.

   V_R = _____ V. Z = _____ Ω.

   V_C = _____ V.   θ = _____ degrees.

   I_T = _____ mA.

   CONCLUSION   Increasing frequency while keeping all other factors the same caused Z to _____, X_C to _____, and θ to _____.

Optional Steps

8. Use the measured and calculated data in the previous steps 2 and 3, and draw a V-I vector diagram in the "Observation" section.

   OBSERVATION

9. Use trigonometry to determine the phase angle.

   OBSERVATION   θ = _____ degrees.

   CONCLUSION   Does the phase angle from the V-I vector diagram agree reasonably with the angle you determined from the Z diagram in step 5? _____.

10. Again, set the source for 100 Hz and a V_A of 3 volts. Use a dual-trace oscilloscope and perform a phase comparison of V_A and circuit current (represented by the voltage across the resistor).

    ! CAUTION: Be sure the signal source ground and the scope ground(s) are connected to the same end (the bottom end in the diagram) of the resistor when making the measurements to prevent the grounds from shorting out a portion of the circuit!

    Determine the phase difference between the two signals.

⚠ **OBSERVATION** θ determined by the scope phase comparison:

θ = _____ degrees.

⚠ **CONCLUSION** Do the scope phase measurements and the phase angle calculations agree reasonably with your earlier findings? (Consider tolerances in components, source and scope frequency calibration tolerances, etc.) _____.

RC Circuits in AC
V, I, R, Z, and θ Relationships in a Parallel RC Circuit

PROJECT 53

Name: _____   Date: _____

FIGURE 53-1

PROJECT PURPOSE To demonstrate the key electrical parameter relationships in a parallel RC circuit. To observe that due to out-of-phase elements, simple dc analysis techniques cannot be used to determine circuit parameters in ac circuits containing reactive components. To compare results of computing circuit parameters using Ohm's law and the Pythagorean theorem.

PARTS NEEDED
- ☐ DMM
- ☐ Function generator or audio oscillator
- ☐ CIS
- ☐ Capacitor 0.1 µF
- ☐ Resistors 100 Ω 10 kΩ

SPECIAL NOTE:

By way of review, recall that in a purely resistive parallel ac circuit, the following conditions exist:

$Z = R_T$; $\theta = 0$ degrees; $I_T =$ arithmetic sum of branch currents; and branch voltage is in phase with branch current for a purely resistive circuit.

In this project, you will examine parameters in a parallel RC circuit. It should be noted again that for this project we will be using the 100-Ω resistor in series with the main line as a circuit current indicator. (Current through R_1 equals ten times its voltage drop.)

PROCEDURE

1. Connect the initial circuit as shown in Figure 53-1.

2. Set the audio oscillator to a frequency of 500 Hz and V_A to 3 volts. Measure the circuit voltages.

   ⚠ OBSERVATION $V_A =$ _____ V. $V_{R_2} =$ _____ V.

   $V_{R_1} =$ _____ V. $V_C =$ _____ V.

   ⚠ CONCLUSION Does the addition of the voltages around any closed loop equal V_A? _____. We conclude from this that V_R and V_C are (*in phase, out of phase*) _____.

3. Calculate I_T from V_{R_1}. Calculate I_{R_2} by Ohm's law. Calculate X_C by using the X_C formula, then calculate I_C using the formula:

   $I_C = V_C / X_C$

   Also, calculate θ using the arctan of I_C / I_R.

   ⚠ OBSERVATION $I_T =$ _____ mA. $I_C =$ _____ mA.

   $I_{R_2} =$ _____ mA.   θ = _____ degrees.

   $X_C =$ _____ Ω.

   ⚠ CONCLUSION Does the total current equal the arithmetic sum of the branch currents? _____.
   This is because the branch currents are _____
   _____ _____. Is the current through R_2 in phase with V_2? _____. The current through the capacitor branch (*leads, lags*) _____ the voltage across the capacitor by close to _____ degrees.
   Since I_T is the vector resultant of I_R and I_C, we would logically conclude that I_T (*leads, lags*) _____ V_A by an angle that is between _____ and _____ degrees.

4. Calculate the circuit Z_T from V_T / I_T.

   ⚠ OBSERVATION $Z_T =$ _____ Ω.

   ⚠ CONCLUSION It is interesting to note that the impedance of a circuit consisting of two parallel 10-kΩ branches, whose currents are 90 degrees out of phase, would ≅ 7.07 kΩ. Does our demonstration circuit result compare closely to this? _____. Why or why not?
   _____
   _____
   _____.

5. Use I_{R_2} and I_C values and determine I_T by means of the Pythagorean theorem.

   ⚠ OBSERVATION I_T calculated = _____ mA.

6. Repeat the previous steps 2 through 5. This time set $f = 250$ Hz and keep V_A at 3 volts.

⚠ OBSERVATION

$V_A =$ _____ V. $I_{R_2} =$ _____ mA.

$V_{R_1} =$ _____ V. $X_C =$ _____ Ω.

$V_{R_2} =$ _____ V. $I_C =$ _____ mA.

$V_C =$ _____ V. $Z_T =$ _____ Ω.

$I_T =$ _____ mA. $\theta =$ _____ degrees.

⚠ CONCLUSION Decreasing frequency caused Z to _____, X_C to _____, and θ to _____.

Story Behind the Numbers
RC Circuits in AC

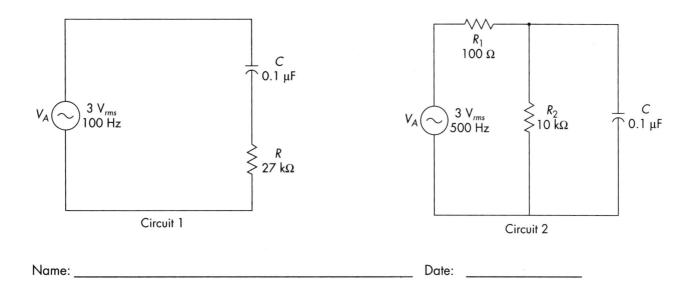

Circuit 1

Circuit 2

Name: _____ Date: _____

Procedure

1. Connect the circuit shown in Circuit 1.

2. Set the frequency of the function generator to 100 Hz and set V_A to 3 volts.

3. Measure and record the values of V_A, V_R, and V_C in the Data Table, as appropriate.

4. From the measured value of V_R, calculate the value of I_T. Calculate the value of X_C, using V_C/I. Calculate the value of circuit impedance (Z) using V_T/I_T. Record this data in the Data Table.

5. Use the Pythagorean approach and calculate and record the value of Z again, as defined in the Data Table.

6. Use trigonometry to determine the circuit phase angle and record in the Data Table.

7. Change the circuit signal frequency input to 200 Hz, still maintaining V_A at 3 volts.

8. At this new frequency, repeat the measurements and calculations performed in steps 1 through 6. Record these new values in the appropriate locations in the Data Table.

9. Reset the source frequency to 100 Hz, still maintaining V_A at 3 volts.

10. Use a dual-trace oscilloscope and perform a phase comparison of V_A and circuit current (represented by the voltage across the resistor).

> **! CAUTION:** Be sure the signal source ground and the scope ground(s) are connected to the bottom end of the resistor (as shown in the diagram) when making the measurements. This will prevent the source and scope grounds from shorting out a portion of the circuit!
>
> Use the appropriate interpretation technique and determine the phase difference between V_A and I_T. If possible, demonstrate your scope patterns to the instructor.

11. Disconnect Circuit 1 and connect the circuit shown in the Circuit 2 diagram.

12. Set the signal source to a frequency of 500 Hz and V_A to 3 volts.

13. Measure the values of V_A, V_{R_1}, V_{R_2}, and V_C and record the results in the Data Table, as appropriate.

14. Calculate the value of I_T using the voltage drop across R_1 and Ohm's law. Calculate the value of X_C using the X_C formula. Calculate the value of I_C by using $I_C = V_C/X_C$. Calculate the phase angle by using the arctan of I_C/I_R. Record all these values in the Data Table, as indicated.

15. Use the Ohm's law approach to find Z_T. Record, as appropriate.

16. Use the Pythagorean approach and determine the value of circuit I_T using the calculated value of I_2 for I_R and the previously computed value of I_C for the I_C in the Pythagorean formula. Record the I_T value in the Data Table, as appropriate.

17. Set the signal source to a frequency of 250 Hz, keeping V_A at 3 volts.

18. Repeat all the measurements and calculations performed in steps 11 through 15 and record results in the appropriate locations in the Data Table.

19. After completing the Data Table, answer the Analysis Questions and produce the brief Technical Lab Report to complete the project.

Analysis Questions

NOTE Answers to these Analysis Questions should be clearly numbered and documented on separate sheets of paper with your name and the date at the top of each page. These answer sheets are to be turned in with the rest of the project documentation, as appropriate.

1. For Circuit 1, at either of the two frequencies used, did the arithmetic sum of the resistor voltage and capacitor voltage equal applied voltage? Explain.

2. For Circuit 1, does the circuit impedance equal the arithmetic sum of resistance and capacitive reactance? Is the impedance greater or less than the sum? Explain.

3. For Circuit 1, which, if any, voltage drops are in phase with the circuit current? Which voltage drops are out of phase with the circuit current? By how much?

Data Table

SERIES RC, CIRCUIT 1				
Component and Parameter I.D.s	**Measured Values @ 100 Hz**	**Calculated Values @ 100 Hz**	**Measured Values @ 200 Hz**	**Calculated Values @ 200 Hz**
V_A (V)		—		—
V_R (V)		—		—
V_C (V)		—		—
I_T (mA)	—		—	
X_C (kΩ)	—		—	
Z, Ohm's law (kΩ)	—		—	
Z, Pythagorean (kΩ)	—		—	
Phase angle (trig)	—			

PARALLEL RC, CIRCUIT 2				
Component and Parameter I.D.s	**Measured Values @ 500 Hz**	**Calculated Values @ 500 Hz**	**Measured Values @ 250 Hz**	**Calculated Values @ 250 Hz**
V_A (V)		—		—
V_{R_1} (V)		—		—
V_{R_2} (V)		—		—
V_C (V)		—		—
I_T, Ohm's law (mA)	—		—	
I_2 (mA)	—		—	
X_C (kΩ)	—		—	
I_C (mA)	—		—	
Z_T, Ohm's law (kΩ)	—		—	
Phase angle (trig)			—	
I_T, Pythagorean (mA)	—			

4. For Circuit 1, what is the phase differential between circuit current and applied voltage called? For this circuit, what was this value when using 100 Hz as the source frequency? What was this value when using the circuit with the 200-Hz signal? Explain why there was a difference. Be specific.

5. When frequency was increased from 100 Hz to 200 Hz, did the circuit impedance increase or decrease? Did it either double or halve? Explain.

6. Did the circuit phase angle change? Was it greater or less at the 200-Hz frequency than at the 100-Hz frequency? Explain why this is true.

7. When using the scope to determine phase angle in step 10 of the procedure steps, were the results reasonably close to your mathematically determined phase angle?

8. For Circuit 2, if you added the voltage around a single "closed loop," did the sum equal the applied voltage? Explain.

9. For Circuit 2, if you arithmetically added the calculated branch currents, did they equal total circuit current? Explain.

10. For Circuit 2, is the current through the capacitor branch in phase with the voltage across the capacitor? If not, explain the relationship of these two parameters.

11. For Circuit 2, does the circuit current lead or lag the circuit applied voltage? Explain.

12. For Circuit 2, was the circuit phase angle greater at 500 Hz or 250 Hz? Explain why this is true.

13. Make a logical comparison between this fact and what was true when the frequency was changed for the series RC circuit (Circuit 1).

14. For Circuit 2, can you calculate circuit impedance using the product-over-the-sum approach? Explain.

15. In summary, answer the following questions with "I" for increase, "D" for decrease, or "RTS" for remain the same.

    For the series RC circuit: A decrease in frequency would cause Z to _____, X_C to _____, and θ to _____.

    For the parallel RC circuit: A decrease in frequency would cause Z to _____, X_C to _____, and θ to _____.

    For a series RC circuit: As X_C increases, circuit phase angle will _____.

    For a parallel RC circuit: As X_C increases, circuit phase angle will _____.

Technical Lab Report

Write a brief technical lab report summarizing the technical facts learned from this project. The report should be organized to provide the following:

1. An introductory paragraph describing the type of circuit being analyzed and the key parameters that will be discussed relating to this circuit.

2. A section describing the most important characteristics of this type of circuit that were shown via the collected data in the tables and graphs.

3. Any special facts or characteristics about this type of circuit that were highlighted in answering the Analysis Questions.

4. A practical example of how the information learned in this project might help you in operating, troubleshooting, error analysis, or adjusting a circuit of this type in your home setting, in your training program setting, or in a job setting in the real world.

5. A summary statement listing the most positive aspects of the project and any parts of the project that were difficult because of equipment problems or unclear instructions. Include areas that might be improved.

PART 15

Summary
RC Circuits in AC

Name: _____ Date: _____

Complete the following review questions, indicating the appropriate response by placing a check in the box next to the correct answer.

1. The higher the frequency of V_A applied to any RC circuit, either series or parallel, the _____ the circuit impedance will be
   - ☐ higher
   - ☐ lower

2. In a purely capacitive series circuit, the arithmetic sum of the individual voltage drops equals V_A.
   - ☐ True
   - ☐ False

3. In a series RC circuit, the arithmetic sum of the individual voltage drops equals V_A.
   - ☐ True
   - ☐ False

4. In a parallel RC circuit, the arithmetic sum of the branch currents equals I_T.
   - ☐ True
   - ☐ False

5. In a series RC circuit, if frequency, resistance, or capacitance increases, the circuit phase angle will
   - ☐ increase
   - ☐ decrease
   - ☐ remain the same

6. In a parallel RC circuit, if frequency, resistance, or capacitance increases, the circuit phase angle will
   - ☐ increase
   - ☐ decrease
   - ☐ remain the same

7. As the resistance in a parallel RC circuit is increased, the circuit will become more
   - ☐ resistive
   - ☐ capacitive

8. As the capacitance in a series RC circuit is increased, the circuit will become more
   - ☐ resistive
   - ☐ capacitive

9. A circuit that is capacitive is one in which the circuit voltage is lagging the circuit current, and I_C and V_C are 90 degrees out of phase.
   ☐ True
   ☐ False

10. The impedance of a parallel RC circuit will increase as frequency is decreased.
    ☐ True
    ☐ False

SERIES RESONANCE

PART 16

Objectives

You will connect several ac *RLC* circuits that illustrate the important characteristics of series resonant circuits.

In completing these projects, you will connect circuits, make measurements, perform calculations, draw conclusions, and be able to answer questions about the following items related to series resonant circuits:

- Definition of resonance
- Relationships of V, I, R, Z, θ, Q, bandwidth, and frequency in series *RLC* circuits
- Finding a circuit's resonant frequency
- Outstanding characteristics of a circuit at series resonance

Project/Topic Correlation Information

PROJECT	TEXT CHAPTER	SECTION	RELATED TEXT TOPIC(S)
54 X_L and X_C Relationships to Frequency	21	21-1	X_L, X_C, and Frequency
55 V, I, R, Z, and θ Relationships when $X_L = X_C$	21	21-2	Series Resonance Characteristics
56 Q and Voltage in a Series Resonant Circuit	21	21-5	Q and Resonant Rise of Voltage
57 Bandwidth Related to Q	21	21-10	Selectivity, Bandwidth, and Bandpass

Series Resonance
X_L and X_C Relationships to Frequency

PROJECT 54

Name: _____   Date: _____

FIGURE 54-1

*Mfg. rated value under specified conditions

PROJECT PURPOSE To verify the opposite effects on X_L and X_C when the frequency of signal applied to a circuit containing both is changed.

PARTS NEEDED
- ☐ DMM
- ☐ Function generator or audio oscillator
- ☐ CIS
- ☐ Capacitor 0.1 µF
- ☐ Inductors 1.5 H, 95 Ω (or approximate)
- ☐ 100 mH
- ☐ Resistor 100 Ω

PROCEDURE

1. Connect the initial circuit as shown in Figure 54-1.

2. Set the function generator to the frequency indicated for the circuit option you are using. Set V_A to 3 volts. Measure V_R and calculate I_T. Measure V_L and V_C and calculate X_L and X_C by Ohm's law.

⚠ OBSERVATION

	Lower f		Higher f	
$V_A =$	_____	V.	$V_A =$ _____	V.
$V_R =$	_____	V.	$V_R =$ _____	V.
$I_T =$	_____	mA.	$I_T =$ _____	mA.
$V_L =$	_____	V.	$V_L =$ _____	V.
$V_C =$	_____	V.	$V_C =$ _____	V.
$X_L =$	_____	Ω.	$X_L =$ _____	Ω.
$X_C =$	_____	Ω.	$X_C =$ _____	Ω.

PART 16: Series Resonance

⚠ CONCLUSION   If the X_C formula was used, the X_C of a 0.1-μF capacitor at 150 Hz would be calculated as _____ ohms for the lower f circuit. For the higher f circuit, the X_C at 1 kHz would be calculated as _____ ohms. Can the difference between the Ohm's law results and the X_C formula be attributed to meter loading effects, capacitor tolerance, and audio oscillator dial calibration tolerance? _____.

3. Change the function generator frequency to 300 Hz for the lower-frequency circuit or to 2 kHz for the higher-frequency circuit. Keep V_A at 3 volts. Measure V_R and calculate I_T. Measure V_L and V_C and calculate X_L and X_C by Ohm's law.

⚠ OBSERVATION

Lower f Higher f

V_A = _____ V. V_A = _____ V.

V_R = _____ V. V_R = _____ V.

I_T = _____ mA. I_T = _____ mA.

V_L = _____ V. V_L = _____ V.

V_C = _____ V. V_C = _____ V.

X_L = _____ Ω. X_L = _____ Ω.

X_C = _____ Ω. X_C = _____ Ω.

⚠ CONCLUSION   When frequency was doubled, the X_L (*increased, decreased*) _____ by approximately ____ times, and the X_C (*increased, decreased*) _____ by about ____ times. We conclude from these results that X_L is (*directly, inversely*) _____ proportional to frequency, and X_C is (*directly, inversely*) _____ proportional to frequency. In a circuit that has both inductance and capacitance, there must be some frequency where X_L and X_C would be _____. In our circuit, would the frequency where this would occur have to be higher or lower than the present settings? _____.

Series Resonance
V, I, R, Z, and θ Relationships When $X_L = X_C$

PROJECT 55

Name: _____   Date: _____

FIGURE 55-1

Lower-Frequency Circuit — V_A 3 V, L 1.5 H*, C 0.1 μF, R 100 Ω

Higher-Frequency Circuit — V_A 3 V, L 100 mH, C 0.1 μF, R 100 Ω

*Mfg. rated value under specified conditions

PROJECT PURPOSE To demonstrate the resonance effects at a frequency where $X_L = X_C$. To show that the series *RLC* circuit basically acts resistively at resonance, inductively above resonance, and capacitively below resonance.

PARTS NEEDED
- ☐ DMM
- ☐ Function generator or audio oscillator
- ☐ CIS
- ☐ Capacitor 0.1 μF
- ☐ Inductors 1.5 H, 95 Ω (or approximate)
- ☐ 100 mH
- ☐ Resistor 100 Ω

PROCEDURE

1. Connect the initial circuit as shown in Figure 55-1.

2. Set V_A at 3 volts and adjust the function generator frequency while monitoring V_R until V_R is maximum. Measure V_R and calculate I_T. Measure V_L and V_C. Calculate X_L and X_C by Ohm's law.

⚠ **OBSERVATION**

Lower *f*

$V_A =$ _____ V.

Frequency = _____ Hz.
(approximately)

Higher *f*

$V_A =$ _____ V.

Frequency = _____ Hz.
(approximately)

PART 16: Series Resonance

Lower f	Higher f
$V_R = $ _____ V.	$V_R = $ _____ V.
$I_T = $ _____ mA.	$I_T = $ _____ mA.
$V_L = $ _____ V.	$V_L = $ _____ V.
$V_C = $ _____ V.	$V_C = $ _____ V.
$X_L = $ _____ Ω.	$X_L = $ _____ Ω.
$X_C = $ _____ Ω.	$X_C = $ _____ Ω.

⚠ **CONCLUSION**   Are X_L and X_C close to being equal at the frequency of maximum V_R? _____.
Would they be equal if circuit and measurement conditions were perfect? _____.
The voltage across the coil (*leads, lags*) _____ the current by about _____ degrees; whereas, the voltage across the capacitor (*leads, lags*) _____ the current by close to _____ degrees. This means that V_L and V_C are close to _____ degrees out of phase with each other. Since I is the same throughout the series circuit and the $I \times X_L$ drop is essentially _____ degrees out of phase with the $I \times X_C$ drop, we may conclude that X_L and X_C reactances are opposite (vectorially) and at resonance are equal. They, therefore, cancel each other's effects on Z. Notice that V_L and V_C both are greater than V_A. This is due to the cancelling effect of the reactances. If X_L and X_C perfectly cancelled (as would be true with perfectly equal and opposite reactances), the circuit Z would equal _____. The resultant circuit current would be (*in phase, out of phase*) _____ with V_A, and the circuit would be acting purely _____.

3. Change the frequency to a new frequency that is well above resonance. Measure V_L and V_C and determine whether the circuit is now acting inductively or capacitively.

   ⚠ **OBSERVATION** V_L is now _____ than V_C.

   ⚠ **CONCLUSION**   The circuit is now acting like an (*RL, RC*) _____ circuit, since X _____ is more than cancelling out X _____. Thus, the circuit is acting like R in series with a resultant X _____. The value of this resultant X is equal to a value of X _____ total minus X _____.
   The result is that I_T (*leads, lags*) _____ V_A by an angle between _____ and _____ degrees; thus the circuit is (*inductive, capacitive*) _____.

4. Keep the same circuit as shown in Figure 55-1.

5. Change the input frequency to a new frequency that is well below the resonant frequency determined in the earlier steps. Measure V_L and V_C and determine whether the circuit is now acting inductively or capacitively.

   ⚠ **OBSERVATION**   New frequency is lower than _____.
   V_L is now _____ than V_C.

⚠ CONCLUSION Since V _____ is now greater than V _____ and the current is the same through both reactances, it indicates that X _____ is greater than X _____. This means that X _____ is more than cancelling X _____ and the circuit is equivalent to an (*RL, RC*) _____ circuit. The circuit impedance therefore would equal the vector resultant of R and a series X _____ whose value equals the difference between _____ and _____. We may conclude from the preceding that at a frequency where $X_L = X_C$ (resonance), the circuit acts essentially like a pure _____ circuit. At frequencies below resonance, a series *RLC* circuit will act equivalent to a simple R _____ circuit. At frequencies above resonance, a *series RLC* circuit will act equivalent to a simple R _____ circuit. We may also conclude that the larger the value of C or L, the (*lower, higher*) _____ will be the resonant frequency and vice versa. Summarizing our observations for a series resonant circuit, we conclude that at resonance Z is (*minimum, maximum*) _____ since X_L and X_C cancel.

Series Resonance
Q and Voltage in a Series Resonant Circuit

Name: _____ Date: _____

FIGURE 56-1

Lower-Frequency Circuit — V_A 3 V, L 1.5 H (rating), C 0.1 µF

Higher-Frequency Circuit — V_A 3 V, L 100 mH (rating), C 0.1 µF

PROJECT PURPOSE To show, via circuit measurements, that the voltage across reactive components in a series resonant circuit is greater than V applied. To demonstrate that the amount of this voltage across each reactive component is related to the circuit Q factor.

PARTS NEEDED
- ☐ DMM
- ☐ Function generator or audio oscillator
- ☐ CIS
- ☐ Capacitor 0.1 µF
- ☐ Inductors 1.5 H, 95 Ω (or approximate)
- ☐ 100 mH

SPECIAL NOTE:

The figure of merit for a resonant circuit is called "Q." In general terms, the higher the ratio of reactance at resonance to series resistance, the higher the Q. If we had a perfect inductor and capacitor tuned to resonance and there was zero resistance in the circuit, the circuit impedance would be zero. Thus, current would be infinite. However, this is impossible in practical circuits. Therefore, Q is considered to be the relationship X_L/R, or X_C/R for series resonant circuits. Also, because the canceling effect of X_L and X_C may allow high circuit current at resonance, there is a voltage magnification across the reactive components at resonance. During this project we will illustrate some of these facts.

PROCEDURE

1. Connect the initial circuit as shown in Figure 56-1.

2. Set V_A at 3 volts and adjust the frequency, while monitoring V_L, for maximum V_L. Measure V_A and V_L and calculate the circuit Q from the ratio of V_L to V_A.

   ⚠ OBSERVATION

   Lower f

   V_A = _____ V.

   V_L = _____ V.
   (approximately)

   Q = _____.
   (approximately)

   f = _____.

   Higher f

   V_A = _____ V.

   V_L = _____ V.
   (approximately)

   Q = _____.
   (approximately)

   f = _____.

   ⚠ CONCLUSION The ratio of X_L/R for this circuit must be approximately _____
   _____.

3. With the same V_A and frequency as the previous step, measure V_C and calculate the circuit Q from V_C/V_A.

   ⚠ OBSERVATION

   Lower f

   V_A = _____ V.

   V_C = _____ V.
   (approximately)

   Q = _____.
   (approximately)

   f = _____.

   Higher f

   V_A = _____ V.

   V_C = _____ V.
   (approximately)

   Q = _____.
   (approximately)

   f = _____.

   ⚠ CONCLUSION The ratio of X_C/R for this circuit must be approximately _____.
   What is the X_C of the capacitor according to the X_C formula? Approximately _____. From the preceding, we may conclude that V_L or V_C equals _____ times V_A; that Q equals the ratio of _____ or _____ to R; and the higher the Q, the (*higher, lower*) _____ the circuit current will be at resonance.

Optional Steps

4. Assuming that the 0.1-μF capacitor is very close to its rated value, determine what the resonant frequency of the circuit would be if the initial circuit inductor was acting at its rated value.

   ⚠ OBSERVATION If the inductor really was acting as rated, the resonant frequency would be approximately

   _____ Hz.   _____ Hz.
   (Lower f)   (Higher f)

5. Assuming a C of 0.1 µF and based on the measured resonant frequency in step 2, how much apparent inductance does this inductor have under the operating conditions used?

⚠ OBSERVATION   "Apparent" inductance is approximately:

_____ H.   _____ mH.
(Lower f)   (Higher f)

⚠ CONCLUSION   It may be concluded that for a given capacitance value, the higher the inductance value used in conjunction with the capacitor, the (*higher, lower*) _____ the resonant frequency will be.

6. Perform appropriate calculations and compare the Q calculated from data in step 2 of Project 55 to the Q determined in step 2 of this project (Project 56).

⚠ OBSERVATION Q calculated from step 2 of Project 55 =

_____ .   _____ .
(Lower f)   (Higher f)

Q calculated from step 2 of this project =

_____ .   _____ .
(Lower f)   (Higher f)

⚠ CONCLUSION   Since the same values of L, C, and V_A are used in both projects, what accounts for the difference in Q values? _____

_____

_____ .

Series Resonance
Bandwidth Related to Q

Name: _____   Date: _____

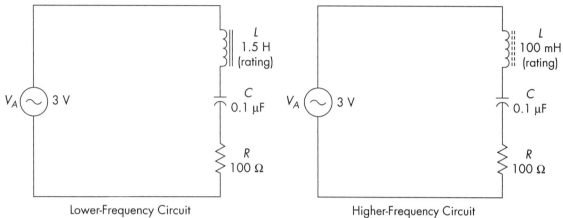

FIGURE 57-1

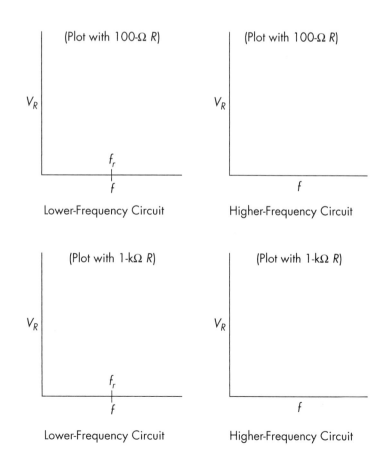

FIGURE 57-2 Sample graph coordinates

PART 16: Series Resonance

NOTE Label f_r and upper and lower bandpass frequencies on each graph. Use separate sheets of graph paper.

PROJECT PURPOSE To provide practical experience and practice in making measurements to find "half-power points" and from these measurements to determine the series *RLC* circuit bandwidth.

PARTS NEEDED
- ☐ DMM
- ☐ Function generator or audio oscillator
- ☐ CIS
- ☐ Capacitor 0.1 µF
- ☐ Inductors 1.5 H, 95 Ω (or approximate)
- ☐ 100 mH
- ☐ Resistors 100 Ω   1 kΩ

SPECIAL NOTE:

Bandwidth for a series resonant circuit may be defined as the difference in frequency between the two frequencies (one below resonance and one above) at which the circuit current is 0.707 (70.7%) of the maximum current (which occurs at resonance). It is interesting to note that the higher the Q of a circuit, the higher the maximum current will be at resonance for any given V_A. If a graph is made of current versus frequency, it will be shown that the higher the Q (and thus the I), the steeper will be the slope of the resonance curve and the smaller the bandwidth. Since Q and bandwidth are related, one formula for bandwidth is: Bandwidth = f_r (resonant frequency)/Q and therefore $Q = f_r$/bandwidth.

PROCEDURE

1. Connect the initial circuit as shown in Figure 57-1.

2. Set V_A to 3 volts. While monitoring V_R, set the frequency of the audio oscillator for resonance (maximum V_R). Measure V_R, V_L, and V_C. Calculate I at resonance.

⚠ **OBSERVATION**

Lower f

$V_A =$ _____ V.
$V_R =$ _____ V.
(approximately)
$V_L =$ _____ V.
(approximately)
$V_C =$ _____ V.
(approximately)
$I =$ _____ mA.
(approximately)

Higher f

$V_A =$ _____ V.
$V_R =$ _____ V.
(approximately)
$V_L =$ _____ V.
(approximately)
$V_C =$ _____ V.
(approximately)
$I =$ _____ mA.
(approximately)

⚠ **CONCLUSION** The approximate Q of this circuit (V_L/V_A): Lower $f =$ _____; Higher $f =$ _____.

According to the formula, Bandwidth = f/Q, the bandwidth of this circuit should be approximately: Lower f circuit = _____ Hz; Higher f circuit = _____ Hz.

If circuit current decreased to 70.7% of I_{max}, the current would be approximately:

Lower f circuit = _____ mA; Higher f circuit = _____ mA. Then, V_R would equal: Lower f circuit = _____ V; Higher f circuit = _____ V.

3. Keeping V_A at 3 volts at each frequency setting, adjust frequency to a frequency below resonance where V_R equals 0.707 of $V_{R_{max}}$. Note this frequency, then change the frequency above resonance until V_R equals 0.707 of $V_{R_{max}}$ and note this frequency. From these two frequencies determine the measured bandwidth.

⚠ **OBSERVATION**   (Lower f)   (Higher f)

Bandwidth = _____ Hz.   Bandwidth = _____ Hz.

f below resonance where V_R is at 70%:

_____ Hz.   _____ Hz.
(Lower f)   (Higher f)

f above resonance where V_R is at 70%:

_____ Hz.   _____ Hz.
(Lower f)   (Higher f)

⚠ **CONCLUSION** Does the measured bandwidth approximate the value calculated from the bandwidth formula in step 2? _____.

4. If time permits, change R to a 1-kΩ resistor and repeat the steps above. Note whether the Q and bandwidth increased or decreased with the higher R value.

⚠ **OBSERVATION**   (Lower f)   (Higher f)

Q = _____. Q = _____.

Bandwidth = _____ Hz.   Bandwidth = _____ Hz.

⚠ **CONCLUSION** The higher the Q, the _____ the bandwidth. The higher the circuit R, the _____ the bandwidth.

5. Make graphic plots of V_R versus f that illustrate bandpass and bandwidth characteristics. Use the data collected in steps 2 through 4 and make your plots on coordinates similar to those shown in Figure 57-2. Use separate graph paper for these plots.

Story Behind the Numbers
Series Resonance

Name: _____   Date: _____

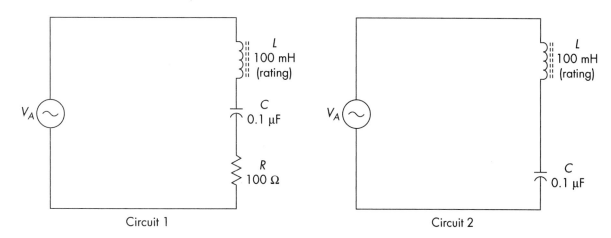

Circuit 1   Circuit 2

NOTE ▶ Prior to performing this project and the upcoming projects that will use inductors, be sure to read the **Special Notes to Students and Instructors** at the beginning of Part 9. For this project, you will use the higher-frequency circuit option; therefore, pay attention to the cautions regarding DMM measuring limitations (frequency limits). If you are using a DMM, be sure it is rated to properly read voltages at the frequencies required in this project. If a scope is to be used, observe the critical ground connection precautions necessary to prevent shorting out components.

Procedure

1. Connect Circuit 1 as shown.

2. Set the signal source to a frequency of 1,000 Hz and V_A to 3 V.

3. Measure V_R and calculate I_T and circuit Z. Record this information in the Data Table, as appropriate.

4. Measure V_L and V_C and calculate the values of X_L and X_C using Ohm's law. Record this data in the Data Table, as indicated.

5. Change the signal source frequency to 2,000 Hz, keeping V_A at 3 V. Make all the measurements and calculations called for in steps 3 and 4. Record your results in the Data Table, as appropriate.

6. With the signal source output still set at 3 V, vary the signal source frequency to higher- and lower-frequency settings while monitoring V_R so as to achieve the resonant frequency setting (maximum voltage across the resistor). While set at the frequency for maximum resistor voltage (resonance), measure V_R and calculate I_T as before. Calculate Z again. Measure V_L and V_C

and calculate X_L and X_C, again using Ohm's law. Record this data in the Data Table.

7. Change the frequency to a new frequency of 3 kHz (a frequency that is above resonant frequency). Measure and record V_L and V_C at this frequency.

8. Change the frequency of signal input to a new frequency of 1 kHz (a frequency that is below the resonant frequency). Measure and record V_L and V_C at this frequency.

9. Keeping V_A at 3 V, make a "frequency run" to enable graphing circuit current versus frequency. (Circuit current is calculated from the value of V_R/R.) Make the frequency run from 1,400 Hz to 2,200 Hz in 200-Hz increments. At each new frequency setting, verify that V_A is 3 V. Measure and record V_R at that frequency to enable calculating circuit current at that frequency.

10. Using the data collected in the preceding step, make a line graph of I_T versus f over the frequency run.

    HINT ▶ If you set up the Excel program worksheet to log your current and frequency data, the "Chart" creation capabilities will help you create this graph!

11. Determine the value that equals 0.707 × current value at resonance (I_{max}). Use this information and the graph to label (1) the approximate resonant frequency, (2) the circuit's bandwidth points, and (3) to identify the approximate bandpass frequencies on the graph. Record these data, as appropriate, in the Data Table.

12. Disconnect the signal source from the circuit and remove the resistor so that the circuit now resembles Circuit 2.

13. Set V_A at 3 V and vary the signal source frequency while monitoring V_L. Adjust the frequency so that V_L is maximum. Reset V_A to 3 V at this frequency (if necessary). Measure V_L and calculate the circuit Q from the ratio of V_L to V_A. Record these measurements and calculation(s) in the Data Table, as appropriate.

14. With the same V_A and frequency as in the preceding step, measure V_C and calculate Q using the ratio of V_C to V_A. Record in the Data Table.

15. **Optional step:** If time permits, you may connect a 1-kΩ resistor in place of the 100-Ω resistor shown in Circuit 1. Vary frequency, as appropriate, to determine what happens to Q and *bandwidth* with the higher resistance in the circuit. Use the V_C/V_A approach for determining Q and use the 0.707 I_{max} points to determine bandwidth.

16. After completing the Data Table, answer the Analysis Questions and develop a brief Technical Lab Report to complete the project.

Data Table

CIRCUIT 1: STEPS 1–8					
Component Parameter I.D.s	Measured Values @ 1,000 Hz	Calculated Values @ 1,000 Hz	Measured Values @ 2,000 Hz	Calculated Values @ 2,000 Hz	
V_A (V)		—		—	
V_R (V)		—		—	
Calculated I_T (mA)	—		—		
Calculated Z (kΩ)	—		—		
V_L (V)		—		—	
V_C (V)		—		—	
Calculated X_L (Ω)	—		—		
Calculated X_C (Ω)	—		—		
V_R (V) @ resonance		—			
I_T (mA) @ resonance	—		—	—	
Calculated Z (Ω) @ resonance	—		—		
V_L (V) @ resonance		—		—	
V_C (V) @ resonance		—		—	
X_L (Ω) @ resonance	—		—	—	
X_C (Ω) @ resonance					
V_L (V) @ above resonance		—	—	—	
V_C (V) @ above resonance		—	—	—	
V_L (V) @ below resonance		—	—	—	
V_C (V) @ below resonance		—	—	—	
CIRCUIT 1: FREQUENCY RUNSTEPS 9–10					
Component Parameter I.D.s	1,400 Hz	1,600 Hz	1,800 Hz	2,000 Hz	2,200 Hz
Value of I_T (V_R/R) = (mA)					

Data Table (continued)

Parameter I.D.s	CIRCUIT 1: *BW, BP* STEP 11				
	Observed Frequencies 100 Ω	Observed Frequencies 1 kΩ			
f @ $0.707 \times I$ at resonance (below resonance)					
f @ $0.707 \times I$ at resonance (above resonance)					
Approx. Bandwidth = _____ (Hz)					
Approx. Bandpass = from _____ to _____ Hz					

Component and Parameter I.D.s	Circuit 2: *Q* Steps 12–14				
	Measured Values with V_L max	Calculated Values with V_L max			
V_A (V)		—			
V_L (max) (V)		—			
Circuit Q (V_L/V_A)	—				
V_C (max) (V)		—			
Circuit Q (V_C/V_A)	—				

Analysis Questions

NOTE Answers to these Analysis Questions should be clearly numbered and documented on separate sheets of paper with your name and the date at the top of each page. These answer sheets are to be turned in with the rest of the project documentation, as appropriate.

1. When frequency was doubled from 1,000 Hz to 2,000 Hz, which reactive parameter approximately doubled and which one decreased to approximately half the original frequency value? Identify which type of reactance is *directly* proportional to frequency and which one is *inversely* proportional to frequency.

2. At the resonant frequency, which component in Circuit 1 was predominant, *C*, *L*, or *R*? Explain.

3. Was the circuit acting inductively, capacitively, or resistively at frequencies above resonance? Does this indicate that the circuit was acting like an *RL* circuit, an *RC* circuit, or like a purely resistive circuit at frequencies above the resonant frequency? Explain.

4. Was the circuit acting inductively, capacitively, or resistively at frequencies below resonance? Does this indicate that the circuit was acting like an *RL* circuit, an *RC* circuit, or like a purely resistive circuit at frequencies below the resonant frequency? Explain.

5. What is the general phase relationship between the voltage across the inductor and the voltage across the capacitor? Is this only true at the resonant frequency? Explain.

6. For Circuit 1 and Circuit 2, does the *circuit* phase angle change as frequency is changed? Explain.

7. From the data in your Data Table, determine whether the circuit Q was higher for Circuit 1 or for Circuit 2. Explain your answer.

8. Indicate if the voltage across the reactive components at the resonant frequency was approximately equal to $Q \times V_A$. What might cause any differences here?

9. At the resonant frequency for both circuits, determine if the calculated values of X_C were approximately the same when using Circuit 1 as when using Circuit 2. List the values in both cases, and explain why this is true.

Technical Lab Report

Write a brief technical lab report summarizing the technical facts learned from this project. The report should be organized to provide the following:

1. An introductory paragraph describing the type of circuit being analyzed and the key parameters that will be discussed relating to this circuit.

2. A section describing the most important characteristics of this type of circuit that were shown via the collected data in the tables and graphs.

3. Any special facts or characteristics about this type of circuit that were highlighted in answering the Analysis Questions.

4. A practical example of how the information learned in this project might help you in operating, troubleshooting, error analysis, or adjusting a circuit of this type in your home setting, in your training program setting, or in a job setting in the real world.

5. A summary statement listing the most positive aspects of the project and any parts of the project that were difficult because of equipment problems or unclear instructions. Include areas that might be improved.

Summary
Series Resonance

Name: _____ Date: _____

Complete the following review questions, indicating the appropriate response by placing a check in the box next to the correct answer.

1. Resonance is sometimes defined as the circuit condition when
   - ☐ $R = X_L$
   - ☐ $R = X_C$
   - ☐ $X_L = Z$
   - ☐ $X_C = Z$
   - ☐ $X_C = X_L$

2. Three circuit factors that determine whether an *RLC* circuit will be equivalent to an *RC*, an *RL*, or a resistive circuit are
   - ☐ V, I, and Z
   - ☐ f, R, and V_A
   - ☐ f, L, and C
   - ☐ none of these

3. If a series *RLC* circuit is at resonance and f is then increased, the circuit will begin to act
   - ☐ capacitively
   - ☐ inductively
   - ☐ resistively

4. What is the phase relationship between V_C and V_L in a series *RLC* circuit?
   - ☐ 0 degrees
   - ☐ 45 degrees
   - ☐ 90 degrees
   - ☐ 180 degrees
   - ☐ none of these

5. What is the phase relationship between I_C and I_L in a series *RLC* circuit?
   - ☐ 0 degrees
   - ☐ 45 degrees
   - ☐ 90 degrees
   - ☐ 180 degrees
   - ☐ none of these

6. As the resonant frequency is approached for any *RLC* circuit, the phase angle will
   - ☐ increase
   - ☐ decrease
   - ☐ remain the same

7. Is it possible for an *RLC* circuit to change from capacitive to inductive simply by changing the frequency of the applied voltage?
   - ☐ Yes
   - ☐ No

8. Impedance in a series *RLC* circuit at resonance is
   - ☐ maximum
   - ☐ minimum
   - ☐ neither of these

9. The voltage across either reactive component in a series resonant RLC circuit is equal to
   - ☐ 0 volts
   - ☐ $R \times C$
   - ☐ $Q \times V_T$
   - ☐ $V_T - V_R$
   - ☐ none of these

10. If the series resistance in a series resonant RLC circuit is decreased, the Q of the circuit will
    - ☐ increase
    - ☐ decrease
    - ☐ remain the same

PARALLEL RESONANCE

Objectives

You will connect several ac *RLC* circuits that illustrate the important characteristics of parallel resonant circuits.

In completing these projects, you will connect circuits, make measurements, perform calculations, draw conclusions, and be able to answer questions about the following items related to parallel resonant circuits:

- Definition of resonance
- Relationships of V, I, R, Z, θ, Q, bandwidth, and frequency in parallel *RLC* circuits
- Finding a circuit's resonant frequency
- Outstanding characteristics of a circuit at parallel resonance

Project/Topic Correlation Information

PROJECT	TEXT CHAPTER	SECTION	RELATED TEXT TOPIC(S)
58 V, I, R, Z, and θ Relationships When $X_L = X_C$	21	21-6	Parallel Resonance Characteristics
59 Q and Impedance in a Parallel Resonant Circuit	21	21-8	Effect of a Coupled Load on the Tuned Circuit
		21-9	Q and the Resonant Rise of Impedance
60 Bandwidth Related to Q	21	21-10	Selectivity, Bandwidth, and Bandpass

Parallel Resonance
V, I, R, Z, and θ Relationships When $X_L = X_C$

PROJECT 58

Name: _____ Date: _____

FIGURE 58-1

Lower-Frequency Circuit | Higher-Frequency Circuit

PROJECT PURPOSE To provide experience in performing analysis of a parallel resonant circuit relative to the key electrical parameters. Additionally, to verify that a parallel resonant circuit acts inductively, below the resonant frequency, and capacitively, above the resonant frequency.

PARTS NEEDED
- ☐ DMM
- ☐ Function generator or audio oscillator
- ☐ CIS
- ☐ Capacitor 0.1 µF
- ☐ Inductors 1.5 H, 95 Ω (or approximate)
- ☐ 100 mH
- ☐ Resistors 100 Ω (2)   1 kΩ

SPECIAL NOTE:

Quickly reviewing, recall that X_L is directly proportional to frequency, and X_C is inversely proportional to frequency. For any given circuit containing L and C, there is a frequency where $X_L = X_C$, often referred to as the *resonant frequency*. Some interesting phenomena occur at resonance, and we will observe some of these for the parallel LC circuit. The Rs used as current indicators do have an effect on the operation of the circuit; however, not to the extent that we cannot observe some of the resonance phenomena.

PROCEDURE

1. Connect the initial circuit as shown in Figure 58-1.

2. Adjust V_A for 3 volts V_{A-B} between points A and B. Monitor V_{R_1} with a voltmeter and adjust the frequency for minimum V_{R_1}. Check that V_{A-B} is still 3 volts. Use the measured V_{R_1} and calculate I_T. Measure V_{R_2} and calculate I_C. Measure V_{R_3} and calculate I_L.

   ⚠ OBSERVATION

	Lower f		Higher f
$V_{A-B} =$	_____ V.	$V_{A-B} =$	_____ V.
$V_{R_1} =$	_____ V.	$V_{R_1} =$	_____ V.
$I_T =$	_____ mA.	$I_T =$	_____ mA.
$V_{R_2} =$	_____ V.	$V_{R_2} =$	_____ V.
$I_C =$	_____ mA.	$I_C =$	_____ mA.
$V_{R_3} =$	_____ V.	$V_{R_3} =$	_____ V.
$I_L =$	_____ mA.	$I_L =$	_____ mA.
$f \cong$	_____ Hz.	$f \cong$	_____ Hz.

   ⚠ CONCLUSION Are I_C and I_L close to being equal at the frequency of minimum V_{R_1}? _____. Would they be precisely equal if we had a pure L and C and ideal measurement conditions? _____. Does I_T equal the sum of the branch currents? _____. This is because I_C and I_L are _____ _____ _____. Since I_C leads V_C by 90 degrees, and I_L lags V_L by about 90 degrees, then I_C and I_L must be about _____ degrees out of phase with each other. Theoretically, if I_C and I_L were exactly equal and 180 degrees out of phase, the resultant I_T would be _____. This means that Z would be _____. In practical circuits, however, Z is maximum at resonance, and its value depends on the Q of the circuit. $Z = Q \times X_L$ (or X_C).

3. Calculate the Z of the parallel resonant circuit using the resultant I_T of the branches as determined from V_{R_1} and the 3 volts V_{A-B} ($Z = V/I$).

   ⚠ OBSERVATION

	Lower f		Higher f
$Z =$	_____ Ω.	$Z =$	_____ Ω.
	(approximately)		(approximately)

   ⚠ CONCLUSION The X_C of the capacitor at the resonant frequency, according to the X_C formula, is about: Lower $f =$ _____ Ω; Higher $f =$ _____ Ω. If $X_L = X_C$, then X_L is about: Lower $f =$ _____ Ω; Higher $f =$ _____ Ω. Notice that Z is greater than either branch's opposition to current. This is in contrast to a purely resistive parallel circuit, where Z is less than the least R branch.

4. Keep the same circuit as shown in Figure 58-1.

5. Change the frequency to a new frequency that is well above f_r. Measure V_{R_2} and V_{R_3}. Calculate I_C and I_L and determine whether the circuit is now acting inductively or capacitively.

⚠ OBSERVATION I_C is now _____ than I_L.

⚠ CONCLUSION Since (I_C, I_L) _____ is now greater than (I_C, I_L) _____, it will cancel it out as far as I_T is concerned. Therefore, the resultant I_T is (*leading, lagging*) _____.

The circuit at this frequency is acting equivalent to a simple R _____ circuit. This means a parallel LC circuit at a frequency that is above resonance will act (*inductively, capacitively*) _____. This is in contrast to a series LC circuit, which above resonance acts _____.

6. Change the input frequency to a new frequency that is well below the resonant frequency of the circuit. Determine I_C and I_L and note whether the circuit is now acting inductively or capacitively.

⚠ OBSERVATION I_C is now _____ than I_L.

⚠ CONCLUSION The circuit is now acting _____, since the predominant current is _____. Thus, we conclude that a parallel LC circuit acts _____ below resonance. This is in contrast to a series LC circuit, which acts _____ below resonance. Summarizing our observations for a parallel resonant circuit (at resonance), we conclude that Z is (*minimum, maximum*) _____, since the branch currents cancel, causing I_T to be (*minimum, maximum*) _____ at resonance. Since the reactive currents cancel, then the resultant circuit current is (*resistive, capacitive, inductive*) _____ and (*in phase, out of phase*) _____ with V_A. This means the circuit phase angle is _____ degrees. In order for the branch currents to completely cancel, they must be _____ degrees out of phase with each other. At resonance, the parallel LC circuit acts (*resistively, capacitively, inductively*) _____; above resonance acts (*resistively, capacitively, inductively*) _____; and below resonance acts (*resistively, capacitively, inductively*) _____.

Parallel Resonance
Q and Impedance in a Parallel Resonant Circuit

PROJECT 59

Name: _____   Date: _____

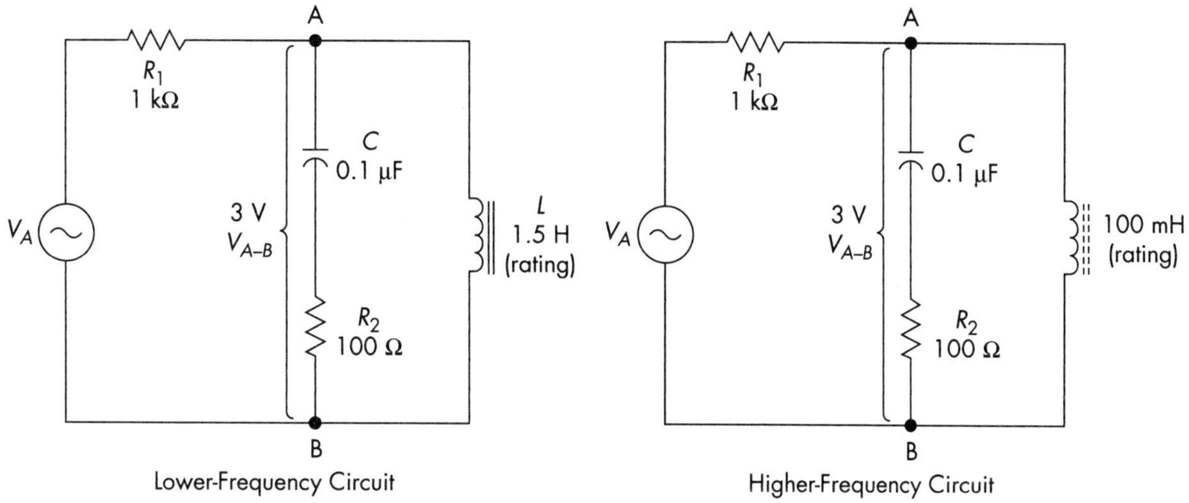

FIGURE 59-1

Lower-Frequency Circuit

Higher-Frequency Circuit

PROJECT PURPOSE To verify that Z is maximum at resonance due to the approximate 180-degree phase relationship of opposite reactance parallel branch currents. To show that the Z of the parallel resonant network is related to the circuit Q.

PARTS NEEDED
- [ ] DMM
- [ ] Function generator or audio oscillator
- [ ] CIS
- [ ] Capacitor
  0.1 µF
- [ ] Inductors
  1.5 H, 95 Ω (or approximate)
- [ ] 100 mH
- [ ] Resistors
  100 Ω
  1 kΩ
  10 kΩ

SPECIAL NOTE:

It should be noted that the Q of a parallel resonant circuit is often considered as the ratio of Z/X_L (or Z/X_C), and that the ratio of branch current to I_T is approximately this same number. For convenience, we will use the ratio of I_C to I_T for determining Q in this project.

PART 17: Parallel Resonance

PROCEDURE

1. Connect the initial circuit as shown in Figure 59-1.

2. Adjust V_A for 3 volts between points A and B. Monitor V_{R_1} and adjust the frequency for minimum V_{R_1} (resonance). Check that V_{A-B} is still 3 volts. Determine I_T and I_C from V_{R_1} and V_{R_2}, respectively.

⚠ OBSERVATION

	Lower f		Higher f
V_{A-B} =	_____ V.	V_{A-B} =	_____ V.
V_{R_1} =	_____ V.	V_{R_1} =	_____ V.
I_T =	_____ mA.	I_T =	_____ mA.
V_{R_2} =	_____ V.	V_{R_2} =	_____ V.
I_C =	_____ mA.	I_C =	_____ mA.

⚠ CONCLUSION I_C is approximately: Lower f circuit = _____; Higher f circuit = _____. _____ times greater than I_T. This indicates a Q of about:
Lower f circuit = _____; Higher f circuit = _____.
The X_C of the capacitor at the resonant frequency, if calculated by the X_C formula, is about: Lower f = circuit _____ Ω; Higher f circuit = _____ Ω.
If the circuit Z is calculated from the formula $Z = Q \times X_C$, the value of Z is: Lower f circuit = _____ Ω; Higher f circuit = _____ Ω. If we calculate Z by Ohm's law using the measured V_A and calculated I_T, the result is Z equals: Lower f circuit = _____ Ω; Higher f circuit = _____ Ω.
Are the two results reasonably comparable considering the measurement errors, and so on? _____.

3. If time permits, connect a 10-kΩ resistor in parallel with points A and B and repeat the procedures of step 2 to determine the new Q and Z. Note whether Q and Z increased or decreased with the shunt 10-kΩ present in the circuit.

⚠ OBSERVATION

	Lower f		Higher f
Q =	_____.	Q =	_____.
Z =	_____ Ω.	Z =	_____ Ω.

⚠ CONCLUSION We may conclude that the lower the R in shunt with a parallel LC circuit, the (*higher, lower*) _____ the Q and Z will be.

Parallel Resonance
Bandwidth Related to Q

PROJECT 60

Name: _____ Date: _____

FIGURE 60-1

Lower-Frequency Circuit   Higher-Frequency Circuit

PROJECT PURPOSE To provide experience in making measurements and performing analysis of a parallel resonant circuit relative to bandwidth and Q. To note how Q and Z are changed when a shunt resistance is introduced to the circuit.

PARTS NEEDED
- ☐ DMM
- ☐ Function generator or audio oscillator
- ☐ CIS
- ☐ Capacitor 0.1 µF
- ☐ Inductors 1.5 H, 95 Ω (or approximate)
- ☐ 100 mH
- ☐ Resistors
  100 Ω
  1 kΩ (2)
  27 kΩ

SPECIAL NOTE:

Bandwidth for a parallel resonant circuit might be defined as the difference in frequency between two frequencies (one below resonance and one above) at which the circuit impedance is 70.7% of the maximum impedance, which occurs at resonance. When the circuit Z is 0.707 of maximum, the circuit current will be 1.414 times I minimum.

319

PROCEDURE

1. Connect the initial circuit as shown in Figure 60-1.

2. Adjust V_A for 3 volts between points A and B. Monitor V_{R_1} and adjust the frequency for minimum V_{R_1} (resonance). Check that V_{A-B} is still 3 volts. Determine I_T and I_C from V_{R_1} and V_{R_2}, respectively. Also determine Q from the formula: $Q = I_C/I_T$.

   ⚠ OBSERVATION

   Lower f

   V_{A-B} = _____ V.
   V_{R_1} = _____ V.
   I_T = _____ mA.
   V_{R_2} = _____ V.
   I_C = _____ mA.
   Q = _____.
   (approximately)

   Higher f

   V_{A-B} = _____ V.
   V_{R_1} = _____ V.
   I_T = _____ mA.
   V_{R_2} = _____ V.
   I_C = _____ mA.
   Q = _____.
   (approximately)

   ⚠ CONCLUSION   Since I_T is the resultant of the two reactive branch currents, we may determine the Z at resonance of the parallel resonant circuit by dividing the 3.0 volts V_{A-B} by I_T. Z for the parallel resonant circuit thus equals approximately _____ Ω for the lower f circuit and _____ Ω for the higher f circuit.

3. Calculate what the current will be at the two frequencies when $Z = 0.707 \times Z_{max}$ by multiplying the I_T at resonance by 1.4.

   ⚠ OBSERVATION I at 70.7% Z points: Lower f = _____ mA; Higher f = _____ mA.

   ⚠ CONCLUSION   If the impedance is 70.7% of Z at resonance, then the current will be 1.4 times I_{min} because 1 divided by 0.707 = _____.

4. Keeping V_{A-B} at 3 volts, find the two frequencies at which Z is 0.707 of Z_{max} by monitoring $I_T (V_{R_1})$. One frequency should be above the resonant frequency, one below. Determine the bandwidth from $f_{hi} - f_{lo}$.

   ⚠ OBSERVATION   Bandwidth equals: Lower f = _____ Hz; Higher f = _____ Hz.

   ⚠ CONCLUSION   If bandwidth is determined from the bandwidth $(\Delta f) = f_r/Q$ formula, using the value of Q from step 2, the value would be approximately: Lower f = _____ Hz; Higher f = _____ Hz. Does the bandwidth in this step reasonably correlate to the $(\Delta f) = f_r/Q$ formula results, all things considered? _____.

5. If time permits, connect a 27-kΩ resistor in shunt with the parallel LC circuit (between points A and B) and repeat the procedures of step 4 to determine bandwidth.

   ⚠ OBSERVATION   The bandwidth with a shunt 27 kΩ is (*more, less*): Lower f = _____; Higher f = _____ than without it.

   ⚠ CONCLUSION   If the bandwidth is greater, the Q of the circuit must have (*increased, decreased*) _____ according to the bandwidth $\Delta f = f_r/Q$ formula.

Story Behind the Numbers
Parallel Resonance

Name: _____ Date: _____

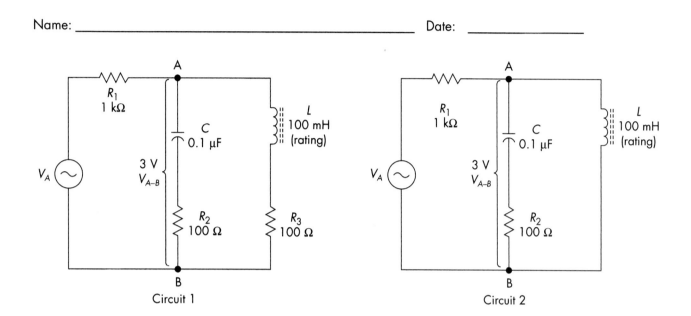

Circuit 1   Circuit 2

NOTE ➤ Prior to performing this project that will use inductors, be sure to read the **Special Notes to Students and Instructors** at the beginning of Part 9. For this project you will use the higher-frequency circuit option; therefore, pay attention to the cautions regarding DMM measuring limitations (frequency limits). If you are using a DMM, be sure it is rated to properly read voltages at the frequencies required in this project. If a scope is to be used, observe the critical ground connection precautions necessary to prevent shorting out components.

Procedure

1. Connect Circuit 1 as shown.

2. Adjust V_A for 3 V between points A and B. Monitor V_1 with a DMM and adjust the source frequency for minimum V_1 value. (This will be the circuit resonant frequency.) Check to see that V_{A-B} is still at 3 V.

3. Measure V_{R_1} and calculate circuit total current. Record the information in the Data Table, as appropriate.

4. Measure V_{R_2} and calculate I_C. Measure V_{R_3} and calculate I_L. Record these values in the Data Table, where indicated.

5. Calculate the value of circuit Z, using the value of voltage across points A and B (3 V) divided by the circuit current (I_T). Record the impedance value in the Data Table, as appropriate.

6. Keep Circuit 1 as shown, but change the frequency to 3 kHz (a new frequency above the resonant frequency). Make sure the voltage across points

A and B is still 3 V. Measure the appropriate circuit parameters to enable calculating the values of I_C and I_L. Record your calculated values of currents in the appropriate locations in the Data Table.

7. Again using Circuit 1, change the frequency to 1 kHz (a new frequency below the resonant frequency). (Keep V_{A-B} at 3 V for this frequency.) Measure the appropriate circuit parameters to enable calculating the values of I_C and I_L. Again, record your calculated values of branch currents in the appropriate locations in the Data Table.

8. Make a frequency run with Circuit 1 in 200-Hz increments from 1,200 Hz to 2,200 Hz. Be sure to reset V_{A-B} to 3 V at each frequency. Measure V_{R_1} and calculate I_T at each frequency setting. Calculate the circuit Z at each frequency. Log the calculated circuit Z in the Data Table for each frequency of the frequency run.

9. Use the collected data to create a line graph showing circuit Z versus frequency.

    HINT ▶ If you set up the Excel program worksheet to log your impedance and frequency data, the "Chart" creation capabilities will help you create this graph!

10. Connect the circuit as shown in the Circuit 2 schematic diagram.

11. With V_{A-B} set to 3 V, adjust the source frequency to achieve resonance (minimum V_{R_1}). Calculate the circuit current. Use V_{A-B} and the calculated circuit current to determine Z at this frequency. Record this value in the Data Table, as indicated.

12. Keeping V_{A-B} at 3 V for each frequency-setting change, adjust the frequency above resonance until I_T is 1.4 times its value at resonance. (This indicates that Z must be 0.707 of its maximum value, which occurred at resonance.) Record this frequency and your calculated value for Z at this frequency in the Data Table.

13. Keeping V_{A-B} at 3 V for each frequency-setting change, adjust the frequency below resonance until I_T is 1.4 times its value at resonance. (Again, this indicates that Z must be 0.707 of its maximum value, which occurred at resonance.) Record this frequency and your calculated value for Z at this frequency in the Data Table.

14. **Optional step:** If time permits, you may connect a 47-kΩ resistor in parallel with the circuit (at points A and B) and perform steps 11 through 13 again. Note what happens to Q and bandwidth as you perform these steps with the parallel resistor added to the circuit.

15. After completing the Data Table, answer the Analysis Questions and develop a brief Technical Lab Report to complete the project.

Data Table

Component and Parameter I.D.s	Measured Values	Calculated Values				
CIRCUIT 1: STEPS 1–9						
$V_{A\text{-}B}$ (V) @ resonance		—				
V_{R_1} (V) @ resonance		—				
Calculated I_T (mA) @ resonance	—					
V_{R_2} (V) @ resonance		—				
Calculated I_C (mA) @ resonance	—					
V_{R_3} (V) @ resonance		—				
Calculated I_L (mA) @ resonance	—					
Calculated Z (kΩ) @ resonance	—					
$V_{A\text{-}B}$ (V) above resonance		—				
V_{R_2} (V) above resonance		—				
Calculated I_C (mA) above resonance	—					
V_{R_3} (V) above resonance		—				
Calculated I_L (mA) above resonance	—					
$V_{A\text{-}B}$ (V) below resonance		—				
V_{R_2} (V) below resonance		—				
Calculated I_C (mA) below resonance	—					
V_{R_3} (V) below resonance		—				
Calculated I_L (mA) below resonance	—					

Data Table (continued)

CIRCUIT 1: FREQUENCY RUN							
Component and Parameter I.D.s	1,200 Hz	1,400 Hz	1,600 Hz	1,800 Hz	2,000 Hz	2,200 Hz	
Calculated Z (kΩ) at each frequency							

CIRCUIT 2: STEPS 10–13						
Component and Parameter I.D.s	Measured Values	Calculated Values	Observed Frequency Values			
V_{A-B} (V) @ resonance		—	—			
Frequency @ resonance		—	—			
Calculated I_T (mA) @ resonance			—			
Calculated Z (kΩ) @ resonance			—			
Calculated $1.4 \times I_T$ (mA) = ?		—	—			
Frequency above resonance for $1.4 \times I_T$		—	—			
Calculated Z (kΩ) @ this frequency			—			
Frequency below resonance for $1.4 \times I_T$		—	—			
Calculated Z (kΩ) @ this frequency			—			

Analysis Questions

NOTE Answers to these Analysis Questions should be clearly numbered and documented on separate sheets of paper with your name and the date at the top of each page. These answer sheets are to be turned in with the rest of the project documentation, as appropriate.

1. At resonance, what causes the circuit current to be minimum?

2. Are the values of currents through the capacitor branch and the inductor branch approximately equal? Explain what you think might cause any differences.

3. Does total current equal the sum of the branch currents? Why is this true?

4. At a frequency above the resonant frequency, did the circuit act inductively or capacitively? Explain.

5. At a frequency below the resonant frequency, did the circuit act inductively or capacitively? Explain.

6. If all components in the circuit were "ideal," how would the circuit act, right at the resonant frequency? Explain.

7. Use the data from your table and determine the X_C of the capacitor at the resonant frequency. Next, determine the Q by using the $Q = Z/X_C$ formula.

8. What general conditions did your graph of Z versus frequency indicate regarding the circuit impedance at resonance, above resonance, and below resonance?

9. Using your data and the graph information, determine and record the bandwidth and bandpass of the circuit on your Analysis Questions answer sheet. Indicate how you determined these values.

10. What would happen to the Circuit Q and bandwidth if a resistor were connected in parallel with the resonant circuit?

Technical Lab Report

Write a brief technical lab report summarizing the technical facts learned from this project. The report should be organized to provide the following:

1. An introductory paragraph describing the type of circuit being analyzed and the key parameters that will be discussed relating to this circuit.

2. A section describing the most important characteristics of this type of circuit that were shown via the collected data in the tables and graphs.

3. Any special facts or characteristics about this type of circuit that were highlighted in answering the Analysis Questions.

4. A practical example of how the information learned in this project might help you in operating, troubleshooting, error analysis, or adjusting a circuit of this type in your home setting, in your training program setting, or in a job setting in the real world.

5. A summary statement listing the most positive aspects of the project and any parts of the project that were difficult because of equipment problems or unclear instructions. Include areas that might be improved.

Summary
Parallel Resonance

Name: _____   Date: _____

Complete the following review questions, indicating the appropriate response by placing a check in the box next to the correct answer.

1. The impedance of a parallel LC circuit at resonance is
   □ maximum   □ neither of these
   □ minimum

2. The resultant total current of a parallel LC circuit at resonance is
   □ maximum   □ neither of these
   □ minimum

3. The current "through" either reactive branch of a parallel LC circuit at resonance is equal to
   □ V_A/Z_T □ f_r/Q
   □ $Q \times X_L$ □ none of these
   □ $Q \times I_T$

4. The impedance of a parallel LC circuit at resonance is equal to
   □ V_A/I_C □ X_L
   □ $Q \times X_L$ □ $X_L + X_C$
   □ X_C □ none of these

5. At a frequency higher than the resonant frequency of a parallel LC circuit, the circuit acts somewhat
   □ inductive   □ capacitive

6. At resonance, the phase angle between V_A and I_T for a parallel LC circuit is
   □ 0 degrees   □ 180 degrees
   □ 45 degrees   □ none of these
   □ 90 degrees

7. If the Q of a parallel resonant circuit is increased, the bandwidth will
   □ increase   □ remain the same
   □ decrease

8. If the Q of a parallel resonant circuit is decreased, the Z will
   □ increase   □ remain the same
   □ decrease

9. At a frequency lower than the resonant frequency, a parallel LC circuit acts somewhat
   □ inductive   □ capacitive

10. If the value of shunt resistance in parallel with a parallel *LC* circuit is changed, will the resonant frequency change?
    ☐ Yes
    ☐ No

THE SEMICONDUCTOR DIODE

PART 18

Objectives

You will connect circuits that illustrate the conduction characteristics of semiconductor diodes and the operation of diode clipper circuits. You will also have a chance to test semiconductor diodes with a VOM ohmmeter and a DMM diode test.

In completing these projects, you will connect circuits, make measurements, perform calculations, draw conclusions, and answer questions about the following items related to semiconductor diodes:

- Forward and reverse bias
- Diode conduction characteristics
- The semiconductor diode symbol
- Typical forward-biased voltage drop
- VOM ohmmeter and DMM tests for semiconductor diodes
- Operation of series and parallel diode clippers

Project/Topic Correlation Information

PROJECT	TEXT CHAPTER	SECTION	RELATED TEXT TOPIC(S)
61 Forward and Reverse Bias, and I vs. V	23	23-1 23-2	Diodes Diode Models
62 Diode Clipper Circuits	23	23-5	Diode Clipper Circuits

The Semiconductor Diode
Forward and Reverse Bias and *I* vs. *V*

PROJECT 61

Name: _____ Date: _____

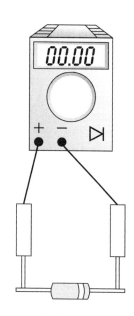

FIGURE 61-1

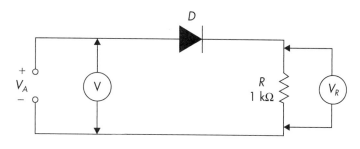

FIGURE 61-2

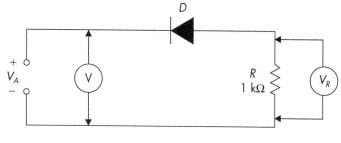

FIGURE 61-3

PROJECT PURPOSE To demonstrate the current-controlling characteristics of a diode when forward biased and reverse biased. A DMM diode checker will be used to test the operating condition of silicon diodes.

PARTS NEEDED
- ☐ DMM (2)
- ☐ VVPS (dc)
- ☐ CIS
- ☐ Diode: Silicon, 1 amp (such as 1N4002)
- ☐ Resistor 1 kΩ

SPECIAL NOTE:

For this project you will first test the diode using a DMM that has a diode testing function. Next, you will use the voltage drop across a 1-kΩ resistor in series with the diode as a means for monitoring the current. Since this is a 1-kΩ resistor, the amount of current in milliamperes is the same as the values you read for the voltage drop across the resistor. Example: If you read 22 V across the 1-kΩ resistor, the current is equal to 22 mA.

Ideally, you will use two meters for this experiment: one for monitoring the value of V_A and the second for measuring the voltages across the resistor and diode.

PROCEDURE

1. Set the DMM to the diode test position. Connect the diode to the DMM with the negative lead to the cathode and the positive lead to the anode. Record the DMM reading.

    ⚠ OBSERVATION   Forward voltage for a good diode: $V_{forward}$ = _____ V.

    ⚠ CONCLUSION   For a good diode, the forward DMM reading is relatively (*high, low*) _____.

2. Switch the leads to the diode, so that the negative lead of the DMM is connected to the anode and the positive lead to the cathode. Record the DMM reading.

    ⚠ OBSERVATION   Reverse voltage for a good diode: $V_{reverse}$ = _____ V.

    ⚠ CONCLUSION   For a good diode, the reverse DMM reading is relatively (*high, low*) _____.

3. Connect the initial circuit as shown in Figure 61-2.

4. Apply 5 volts to the circuit with the anode connected to the positive side of the source. Determine the circuit current from V_R for the forward-biased diode.

    ⚠ OBSERVATION   I = _____ mA. (approximately)

    ⚠ CONCLUSION   Is the diode conducting with the anode positive with respect to the cathode? _____. The voltage drop across the forward-biased diode is approximately _____ V.

5. Turn the diode around so that the positive side of the source will be connected to the cathode of the diode as shown in Figure 61-3.

6. Apply 5 volts to the circuit. Determine the circuit current for V_R for the reverse-biased diode.

⚠ OBSERVATION $I =$ _____ mA.

⚠ CONCLUSION   Is the diode conducting with the cathode positive with respect to the anode? _____. For the reverse-biased condition, the diode is acting like (*an open, a short*) _____ circuit. The resistance of the reverse-biased diode is virtually _____. Since $V_R = 0$ V, the voltage across the diode must equal _____.

7. Turn the diode around once more in order to forward bias it, then increase V_A in small steps, as indicated in the "Observation" section, and note the current at each step.

⚠ OBSERVATION

For $V_A = 0.5$ V: $I =$ _____ mA. $V_A = 6$ V: $I =$ _____ mA.
For $V_A = 1$ V: $I =$ _____ mA. $V_A = 7$ V: $I =$ _____ mA.
$V_A = 2$ V: $I =$ _____ mA. $V_A = 8$ V: $I =$ _____ mA.
$V_A = 3$ V: $I =$ _____ mA. $V_A = 9$ V: $I =$ _____ mA.
$V_A = 4$ V: $I =$ _____ mA. $V_A = 10$ V: $I =$ _____ mA.
$V_A = 5$ V: $I =$ _____ mA.

⚠ CONCLUSION   As V_A was increased, the current (*increased, decreased*) _____. We can deduce, therefore, that diode conduction is (*directly, inversely*) _____ proportional to V_A when forward biased. What is the approximate dc resistance of the diode when V_A is 7 volts? _____ Ω. We conclude from the preceding that a diode will conduct when the anode is (*positive, negative*) _____ with respect to the cathode.

8. With the diode forward biased, measure the voltage drop across the diode for each increase in V_A as shown in the "Observation" section.

⚠ OBSERVATION

$V_A = 0.5$ V: $V_D =$ _____ V. $V_A = 4$ V: $V_D =$ _____ V.
$V_A = 1$ V: $V_D =$ _____ V. $V_A = 10$ V: $V_D =$ _____ V.
$V_A = 2$ V: $V_D =$ _____ V.

⚠ CONCLUSION   As V_A reached a certain level of forward biasing, the voltage drop across the diode (*decreased, increased, stayed about the same*) _____. This implies that the resistance of the diode changed (*linearly, nonlinearly*) _____ as V_A was increased.

The Semiconductor Diode
Diode Clipper Circuits

PROJECT 62

Name: _____ Date: _____

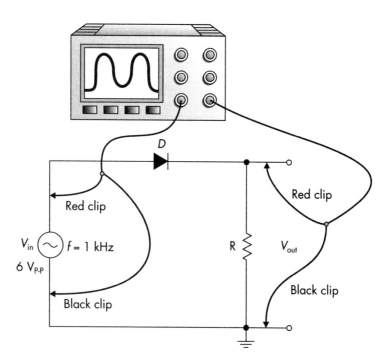

FIGURE 62-1

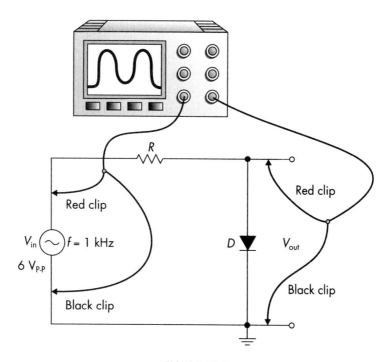

FIGURE 62-2

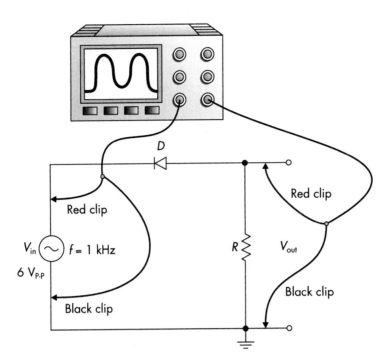

FIGURE 62-3

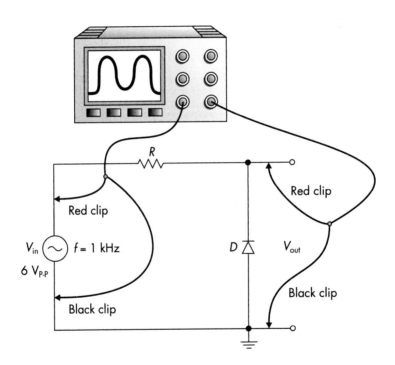

FIGURE 62-4

PROJECT PURPOSE In this project you will observe the actual operation of all four possible types of diode clipper circuits.

PARTS NEEDED
- ☐ Dual-trace oscilloscope
- ☐ Function generator or audio oscillator
- ☐ CIS
- ☐ Diodes: Silicon (2)
- ☐ Resistor 27 kΩ

PROCEDURE

1. Connect the circuit exactly as shown in Figure 62-1.

2. Set the function generator (sine-wave mode) or audio oscillator to a frequency of 1 kHz and an amplitude (V_{in}) of 6 V_{P-P}. Draw the waveform for V_{in} (2 cycles), clearly indicating the positive and negative peak voltage values.

   ⚠ OBSERVATION   Waveform:

   V_{in} positive peak voltage = _____ V.
   V_{in} negative peak voltage = _____ V.

3. Draw the waveform observed at V_{out}. Again, clearly indicate the positive and negative peak voltage values.

   ⚠ OBSERVATION   Waveform:

   V_{out} positive peak voltage = _____ V.
   V_{out} negative peak voltage = _____ V.

   ⚠ CONCLUSION   This circuit is a (*series, parallel*) _____ diode clipper. The output waveform indicates this circuit is a (*positive, negative*) _____ clipper. During the positive half-cycle of V_{in}, the diode is (*forward, reverse*) _____ biased and current (*is, is not*) _____ flowing through the resistor.

4. Connect the circuit exactly as shown in Figure 62-2.

5. Set the function generator (sine-wave mode) or audio oscillator to a frequency of 1 kHz and an amplitude (V_{in}) of 6 V_{P-P}. Sketch the waveform for V_{in} (2 cycles), clearly indicating the positive and negative peak voltage values.

   ⚠ OBSERVATION   Waveform:

   V_{in} positive peak voltage = _____ V.
   V_{in} negative peak voltage = _____ V.

6. Draw the waveform observed at V_{out}, and show the positive and negative peak voltage values.

   OBSERVATION   Waveform:

   V_{out} positive peak voltage = _____ V.
   V_{out} negative peak voltage = _____ V.

   CONCLUSION   This circuit is a (*series, parallel*) _____ diode clipper. The output waveform indicates this circuit is a (*positive, negative*) _____ clipper. During the positive half-cycle of V_{in}, the diode is (*forward, reverse*) _____ biased and current (*is, is not*) _____ flowing through the resistor.

7. Connect the circuit exactly as shown in Figure 62-3.

8. Set the input waveform (V_{in}) as a 6-V_{P-P}, 1-kHz sine wave. Sketch the waveform (2 cycles), clearly indicating the positive and negative peak voltage values.

   OBSERVATION   Waveform:

   V_{out} positive peak voltage = _____ V.
   V_{out} negative peak voltage = _____ V.

9. Sketch the waveform observed at V_{out}, clearly indicating the positive and negative peak voltage values.

   OBSERVATION   Waveform:

   V_{out} positive peak voltage = _____ V.
   V_{out} negative peak voltage = _____ V.

   CONCLUSION   This circuit is a (*series, parallel*) _____ diode clipper. The output waveform indicates this circuit is a (*positive, negative*) _____ clipper. During the positive half-cycle of V_{in}, the diode is (*forward, reverse*) _____ biased and current (*is, is not*) _____ flowing through the resistor.

10. Connect the circuit exactly as shown in Figure 62-4.

11. Set the input waveform (V_{in}) as a 6-V_{P-P}, 1-kHz sine wave. Sketch the waveform (2 cycles), clearly indicating the positive and negative peak voltage values.

    OBSERVATION   Waveform:

    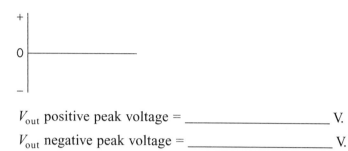

    V_{out} positive peak voltage = _____ V.

    V_{out} negative peak voltage = _____ V.

12. Sketch the waveform observed at V_{out}, clearly indicating the positive and negative peak voltage values.

    OBSERVATION   Waveform:

    V_{out} positive peak voltage = _____ V.

    V_{out} negative peak voltage = _____ V.

    CONCLUSION   This circuit is a (*series, parallel*) _____ diode clipper. The output waveform indicates this circuit is a (*positive, negative*) _____ clipper. During the positive half-cycle of V_{in}, the diode is (*forward, reverse*) _____ biased and current (*is, is not*) _____ flowing through the resistor.

Summary
The Semiconductor Diode

Name: _____ Date: _____

Complete the following review questions, indicating the appropriate response by placing a check in the box next to the correct answer.

1. In analyzing the symbol for the semiconductor diode,
   - ☐ the arrow is the cathode and the "flat bar" is the anode
   - ☐ the arrow is the anode and the "flat bar" is the cathode
   - ☐ neither of these

2. When a diode is forward biased,
   - ☐ the anode is negative with respect to the cathode
   - ☐ the cathode is positive with respect to the anode
   - ☐ the anode is positive with respect to the cathode
   - ☐ none of these

3. A diode in which the cathode is more negative than the anode is
   - ☐ forward biased
   - ☐ neither of these
   - ☐ reverse biased

4. As the forward bias on a diode is increased, the current will
   - ☐ increase
   - ☐ remain the same
   - ☐ decrease

5. A reverse-biased rectifier diode acts like
   - ☐ a short
   - ☐ a 10-kΩ resistor
   - ☐ an open

6. The voltage drop across a forward-conducting silicon diode is approximately
   - ☐ 0.25 V
   - ☐ 1.0 V
   - ☐ 0.5 V
   - ☐ 2.0 V
   - ☐ 0.7 V

7. When testing a good rectifier diode with a DMM in diode test mode and with the positive lead connected to the anode,
   - ☐ the reading is 0
   - ☐ the reading is 0L
   - ☐ the reading is 0.7
   - ☐ none of these

8. When testing a shorted rectifier diode with a DMM in diode test mode and with the positive lead to the anode,
   - ☐ the reading is 0
   - ☐ the reading is 0L
   - ☐ the reading is 0.7
   - ☐ none of these

9. In a series diode clipper circuit, the voltage output is taken across the
   - ☐ power source
   - ☐ diode
   - ☐ resistor

10. Reversing the polarity of the diode in a parallel diode clipper circuit
    ☐ reverses the clipping polarity
    ☐ changes it to a series clipper circuit
    ☐ shorts out the power source
    ☐ has no effect on the circuit

SPECIAL-PURPOSE DIODES

PART 19

Objectives

You will connect circuits that illustrate the special voltage-regulating and voltage-clipping characteristics of zener diodes, and the light-emitting qualities of light-emitting diodes.

In completing these projects, you will connect circuits, make measurements, perform calculations, draw conclusions, and answer questions about the following items related to zener diodes and light-emitting diodes (LEDs):

- Reverse-bias characteristics of zener diodes
- Zener-diode action when the applied voltage is below the rated zener breakdown voltage level
- Zener diode action when the applied voltage is above the rated zener breakdown voltage level
- The zener diode used in a clipping circuit
- Forward-bias characteristics of LEDs
- Values of series-limiting resistors

Project/Topic Correlation Information

PROJECT	TEXT CHAPTER	SECTION	RELATED TEXT TOPIC(S)
63 Zener Diodes	23	23-6	Zener Diodes
64 Light-Emitting Diodes	23	23-7	The Light-Emitting Diode (LED)

Special-Purpose Diodes
Zener Diodes

PROJECT 63

Name: _____ Date: _____

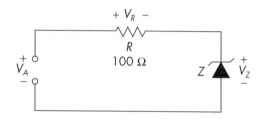

FIGURE 63-1

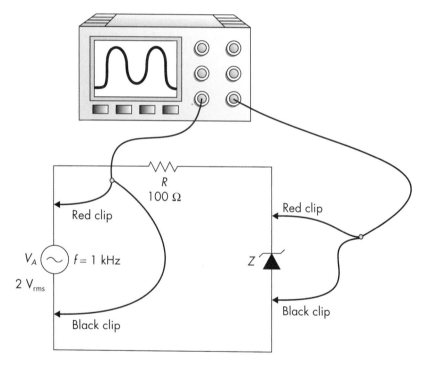

FIGURE 63-2

PROJECT PURPOSE In this project you will observe the regulating and clipping action of a zener diode.

PARTS NEEDED
- ☐ DMM (2)
- ☐ Dual-trace oscilloscope
- ☐ VVPS (dc)
- ☐ CIS
- ☐ Function generator or audio oscillator
- ☐ Zener diode: 5.1 V, 1 W (1N4733 or equivalent)
- ☐ Resistor 100 Ω

345

SPECIAL NOTE:

You can closely approximate the amount of current flowing through the circuit (in mA) by measuring the voltage across the 100-Ω resistor and multiplying the value by ten. Example: If you read 2.2 V across the resistor, the current through the circuit will be close to 22 mA. Ideally, you will use two meters for this experiment: one for monitoring V_A, and the second for taking measurements across the resistor (V_R) and across the diode (V_Z).

PROCEDURE

1. Connect the circuit shown in Figure 63-1.

   ▲ CONCLUSION   The zener diode in this circuit is properly (*forward, reverse*) _____ biased. What is the rated zener voltage (V_{ZT}) for this zener diode? _____ V.

2. Set V_A to 3 volts. Measure the voltage across the zener diode (V_Z) and the voltage across the resistor (V_R), and determine the current through the circuit (I_R). Increase V_A in small steps, as indicated in the "Observation" section, and determine the values of V_Z, V_R, and I_R for each step.

   ▲ OBSERVATION
   For $V_A = 3$ V: $V_Z =$ _____ V, $V_R =$ _____ V, $I_R =$ _____ mA.
   For $V_A = 4$ V: $V_Z =$ _____ V, $V_R =$ _____ V, $I_R =$ _____ mA.
   For $V_A = 5$ V: $V_Z =$ _____ V, $V_R =$ _____ V, $I_R =$ _____ mA.
   For $V_A = 6$ V: $V_Z =$ _____ V, $V_R =$ _____ V, $I_R =$ _____ mA.
   For $V_A = 7$ V: $V_Z =$ _____ V, $V_R =$ _____ V, $I_R =$ _____ mA.

   ▲ CONCLUSION   As long as V_A remains below V_{ZT}, the voltage across the zener diode (*equals V_A, remains close to the diode's V_{ZT} value, remains very close to zero volts*) _____ _____. While V_A is above V_{ZT}, the voltage across the zener diode (*equals V_A, remains close to the diode's V_{ZT} value, remains very close to zero volts*) _____.

   While V_A remains below V_{ZT}, the current through the zener diode (*changes with V_A, remains at a regulated level, remains very close to zero*) _____ _____. Once V_A rises above V_{ZT}, the current through the zener diode (*changes with V_A, remains at a regulated level, remains very close to zero*) _____.

3. Return V_A to 0 volts. Determine the exact zener breakdown voltage (V_{ZT}) by gradually increasing the value of V_A while observing the voltage across the zener diode. Note the values where further increasing V_A no longer causes a significant increase in the zener voltage.

   ▲ OBSERVATION $V_{ZT} =$ _____ V.

   ▲ CONCLUSION   The voltage across a zener diode remains at the fixed V_{ZT} level as long as V_A is (*below, above*) _____ V_{ZT}.

4. Replace the VVPS for V_A with the function generator or audio generator (Figure 63-2). Set the function generator (sine-wave mode) to 1 kHz at 6.3 V_{rms}.

5. Sketch the waveform you find at V_A, indicating the positive and negative peak values.

   ⚠ OBSERVATION   Waveform:

   Positive peak V_A = _____ V.
   Negative peak V_A = _____ V.

6. Sketch the waveform you find across the zener diode. Indicate the peak positive and peak negative voltages.

   ⚠ OBSERVATION   Waveform:

   Positive peak V_Z = _____ V.
   Negative peak V_Z = _____ V.

   ⚠ CONCLUSION   Explain the shape of the waveform that occurs across the zener diode during the positive half-cycle of the input waveform. Account for the shape of the waveform that occurs across the zener diode during the negative half-cycle of the input waveform. Describe how this circuit could be considered a clipping circuit.

   NOTE ➤ Use space below for explanations.

Special-Purpose Diodes
Light-Emitting Diodes

PROJECT 64

Name: _____ Date: _____

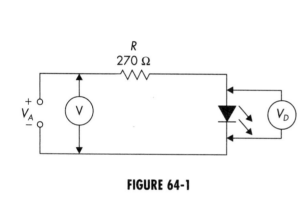

FIGURE 64-1

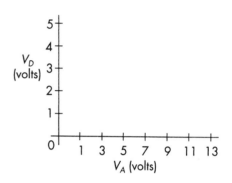

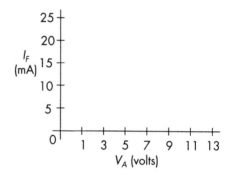

FIGURE 64-2

PROJECT PURPOSE In this experiment you will observe the operation of a typical LED. Ideally, you will use two meters for this experiment: one to monitor the applied dc voltage (V_A) and a second to monitor the forward voltage drop across the LED (V_D).

PARTS NEEDED
- ☐ DMM (2)
- ☐ VVPS (dc)
- ☐ CIS
- ☐ LED: gallium-arsenide red, 20 mA
- ☐ Resistors
  - 47 Ω   270 Ω
  - 220 Ω   330 Ω

SPECIAL NOTE:

You will need the following formula to calculate the amount of forward current through the LED.

$$I_D = (V_A - V_D)/R$$

where:
- I_D is the forward current through the diode
- V_A is the voltage applied to the circuit
- V_D is the forward voltage drop across the LED
- R is the value of the resistor connected in series with the LED

PROCEDURE

1. Connect the circuit as shown in Figure 64-1.

2. Set V_A to 1 V dc. Measure the voltage across the LED, calculate the amount of current flowing through the circuit, and note the intensity of the light (none, dim, moderately bright, bright, very bright). Increase V_A in steps indicated in the "Observation" section. Determine V_D and I_D, and note the brightness level in each case.

   ⚠ OBSERVATION

   $V_A = 1$ V; $V_D = \underline{\quad}$ V, $I_D = \underline{\quad}$ mA, brightness = \underline{\hspace{3cm}}.

   $V_A = 3$ V; $V_D = \underline{\quad}$ V, $I_D = \underline{\quad}$ mA, brightness = \underline{\hspace{3cm}}.

   $V_A = 5$ V; $V_D = \underline{\quad}$ V, $I_D = \underline{\quad}$ mA, brightness = \underline{\hspace{3cm}}.

   $V_A = 7$ V; $V_D = \underline{\quad}$ V, $I_D = \underline{\quad}$ mA, brightness = \underline{\hspace{3cm}}.

   $V_A = 9$ V; $V_D = \underline{\quad}$ V, $I_D = \underline{\quad}$ mA, brightness = \underline{\hspace{3cm}}.

   $V_A = 11$ V; $V_D = \underline{\quad}$ V, $I_D = \underline{\quad}$ mA, brightness = \underline{\hspace{3cm}}.

   $V_A = 13$ V; $V_D = \underline{\quad}$ V, $I_D = \underline{\quad}$ mA, brightness = \underline{\hspace{3cm}}.

   ⚠ CONCLUSION Complete the graphs in Figure 64-2 which show how V_D increases with V_A, and how I_D increases with V_A. The brightness of the LED seems to correlate better with the amount of (V_D, I_D) \underline{\hspace{2cm}} than with the amount of (V_D, I_D) \underline{\hspace{2cm}}. Based on the data you observed, the forward junction potential for this LED is about \underline{\hspace{2cm}} V.

3. Set V_A to 3 V dc, replace R with a 47-Ω resistor, and complete the blanks in the "Observation" section. Repeat the steps for the values of V_A and R indicated.

   ⚠ OBSERVATION

   $V_A = 3$ V; $R = 47$ Ω, $I_D = \underline{\quad}$ mA, brightness = \underline{\hspace{3cm}}.

   $V_A = 6$ V; $R = 220$ Ω, $I_D = \underline{\quad}$ mA, brightness = \underline{\hspace{3cm}}.

   $V_A = 9$ V; $R = 330$ Ω, $I_D = \underline{\quad}$ mA, brightness = \underline{\hspace{3cm}}.

   ⚠ CONCLUSION It is possible to use a given LED with just about any amount of applied dc as long as you select the correct value of series resistor for approximately 20 mA of forward current. (*True, False*) \underline{\hspace{4cm}}.

Summary
Special-Purpose Diodes

Name: _____ Date: _____

Complete the following review questions, indicating the appropriate response by placing a check in the box next to the correct answer.

1. A zener diode is normally connected into a dc circuit in a direction that causes it to be
   ☐ reverse biased   ☐ forward biased

2. When a zener diode is properly connected into a variable-voltage dc circuit, the zener conducts whenever the dc source
   ☐ is greater than zero volts
   ☐ exceeds the zener's forward junction potential
   ☐ exceeds the zener's reverse breakdown voltage

3. A forward-biased zener diode behaves much the same as a forward-biased silicon rectifier diode.
   ☐ True
   ☐ False

4. When a zener diode rated at 12 V is operating from an 18-V source and the series resistor has a value of 100 Ω, the current through the zener diode is approximately
   ☐ 12 mA   ☐ 60 mA
   ☐ 18 mA   ☐ 120 mA

5. Used as a voltage clipper, a zener diode clips one polarity at the forward junction potential and the opposite polarity at the reverse breakdown potential.
   ☐ True
   ☐ False

6. The schematic symbols for an LED and a photodiode
   ☐ are the same except the arrows point in different directions
   ☐ are exactly the same
   ☐ bear no resemblance to one another

7. An LED is connected into a dc circuit in a direction that causes it to
   ☐ be reverse biased
   ☐ gather light energy
   ☐ be forward biased
   ☐ break down and conduct backward

8. The forward junction potential for an LED tends to be _____ that of a rectifier diode.
   ☐ less than   ☐ greater than
   ☐ about the same as

9. The reverse breakdown voltage for an LED tends to be _____ that of a rectifier diode.
    ☐ less than
    ☐ greater than
    ☐ about the same as

10. Suppose you want to operate an LED from a 12-Vdc source, and the known specifications for the LED are $V_D = 1.7$ V when $I_D = 15$ mA. Which of the following practical resistors is closest to the correct value for the series resistor you should use?
    ☐ 270 Ω
    ☐ 330 Ω
    ☐ 470 Ω
    ☐ 680 Ω
    ☐ 1 kΩ

POWER SUPPLIES

PART 20

Objectives

You will connect half-wave and bridge rectifier circuits and observe their operation with and without capacitor filtering. You will also demonstrate the operation of a voltage doubler power supply.

In completing these projects, you will connect circuits, make measurements, perform calculations, draw conclusions, and be able to answer questions about the following items related to power supplies:

- Relationship of output ripple frequency to ac input frequency
- Relationship of output voltage to ac input voltage
- Comparison of filtered and unfiltered outputs
- Distinction between half-wave and full-wave rectification
- Relationship of the amount of output ripple to the amount of capacitor filtering

Project/Topic Correlation Information

PROJECT	TEXT CHAPTER	SECTION	RELATED TEXT TOPIC(S)
65 Half-Wave Rectifier	24	24-2	Half-Wave Rectifier Circuits
		24-3	Capacitance Filters
66 Bridge Rectifier	24	24-2	The Bridge Rectifier
		24-3	Capacitance Filters
67 Voltage Multiplier	24	24-5	Basic Voltage Multiplier Circuits
Voltage Regulator		24-6	Advanced Power Supply Topics

353

Power Supplies
Half-Wave Rectifier

PROJECT 65

Name: _____ Date: _____

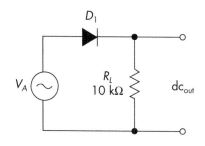

FIGURE 65-1

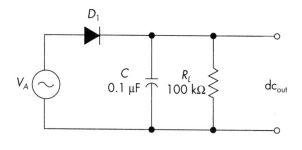

FIGURE 65-2

PROJECT PURPOSE The purpose of this project is to demonstrate the action of a simple half-wave rectifier and the effects of capacitor filtering.

PARTS NEEDED
- ☐ DMM
- ☐ Oscilloscope
- ☐ CIS
- ☐ Function generator or audio oscillator
- ☐ Diode: Silicon, 1-amp rating
- ☐ Resistors
  10 kΩ
  100 kΩ
- ☐ Capacitors
  0.1 μF
  1.0 μF

SPECIAL NOTE:

The following formulas will be helpful for drawing proper conclusions about the results of your work:

Input $V_{rms} = 0.707 \times V_{pk}$ Effective ac

Output $V_{dc} = 0.318 \times V_{pk}$ Average dc out

PROCEDURE

1. Connect the initial circuit as shown in Figure 65-1.

2. Set the function generator (sine-wave mode) or audio oscillator to a frequency of 100 Hz and V_A to 3 V_{rms}. Sketch two complete cycles of the input waveform in the "Observation" section.

⚠ **OBSERVATION**   Waveform:

V_p input = _____ V.

355

3. Measure the dc voltage output across R_L (dc_{out}).

   OBSERVATION dc_{out} = _____ V.

   CONCLUSION   The dc output voltage should be approximately what percentage of the rms input voltage? Approximately _____ %. Since the average voltage value over one ac alternation is $0.637 \times V_p$ and the effective value is $0.707 \times V_{pk}$, then V_{dc} must be about _____ tenths of V_{rms}, because 0.637 is about _____ tenths of 0.707. Since the diode can conduct only half the time with ac applied, the average dc_{out} for the half-cycle the diode *does not* conduct is _____ V.

4. Connect the oscilloscope across R_L. Sketch the waveform for two complete cycles in the "Observation" section. Indicate the maximum and minimum voltage levels.

   OBSERVATION   Waveform:

   CONCLUSION   For the half-cycle the diode does conduct, the peak V_{out} should be about _____ times V_{rms} (neglecting the small diode voltage drop). The end result is that the average dc_{out} over the *entire cycle* of ac input is about _____ times V_{rms} (again, neglecting the diode voltage drop).

5. Connect a 0.1-µF filter capacitor in parallel with a new value of R_L, as shown in Figure 65-2.

6. Set the function generator (sine-wave mode) to a frequency of 100 Hz and an amplitude of 3 V_{rms}. Sketch two complete cycles of the input waveform. Measure and record the ac input voltage (ac_{in}).

   OBSERVATION ac_{in} = _____ V.
   Waveform:

7. Measure and record the dc voltage output across R_L (dc$_{out}$). Sketch two complete ac cycles of the oscilloscope waveform found across R_L.

⚠ **OBSERVATION** dc$_{out}$ = _____ V.

Waveform:

⚠ **CONCLUSION** Is the dc output voltage higher or lower than the effective ac input voltage? _____. This can be explained by the fact that the filter capacitor charges to the _____ _____ value of the input voltage, rather than the average or effective values. Since the charge path for the capacitor is through the low resistance of the forward-conducting diode, the capacitor has time to charge to the _____ _____ value of the input voltage. But when the input voltage begins to decrease from its _____ value, the capacitor begins to _____ slowly through the load resistor, R_L. Before the capacitor can discharge completely, the next cycle of ac will reach a point where it is higher than the voltage remaining on the capacitor. The diode thus begins to conduct again and charge the capacitor to the _____ voltage value again.

8. Replace the 0.1-µF filter capacitor with a 1.0-µF capacitor.

9. Record the dc$_{out}$ and sketch two complete ac cycles of the oscilloscope waveform found across R_L.

⚠ **OBSERVATION** dc$_{out}$ = _____ V.

Waveform:

⚠ **CONCLUSION** Is dc$_{out}$ higher or lower with the 1.0-µF filter capacitor? _____. This is because the *RC* discharge time with the 1.0-µF filter capacitor is much _____ than with the 0.1-µF capacitor. The larger the value of filter capacitor, the (*higher, lower*) _____ the amount of discharge during each nonconducting half-cycle for the diode. This means the average dc output voltage (*increases, decreases*) _____ with increasing values of filter capacitance.

Power Supplies
Bridge Rectifier

PROJECT 66

Name: _____ Date: _____

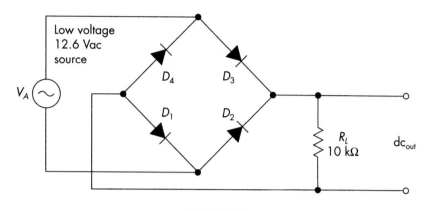

FIGURE 66-1

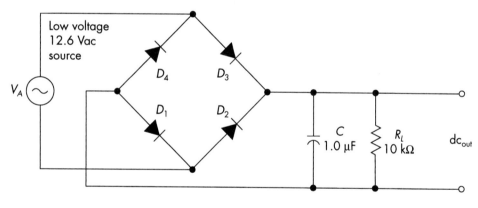

FIGURE 66-2

PROJECT PURPOSE In this project you will study the action of a bridge rectifier and demonstrate the effects of capacitor filtering.

PARTS NEEDED
- ☐ DMM
- ☐ Oscilloscope
- ☐ CIS
- ☐ Function generator or audio oscillator
- ☐ Diodes: Silicon, 1-amp rating (4)
- ☐ Resistor 10 kΩ
- ☐ Capacitor 1.0 μF
- ☐ Transformer (12.6 V with center-tapped secondary) or, a 12.6 Vac source

PROCEDURE

1. Connect the initial circuit as shown in Figure 66-1.

2. Apply 12.6 Vac to the circuit. Measure the dc voltage output across R_L (dc_{out}).

3. Sketch the secondary waveform. Indicate the maximum and minimum voltage levels. Measure the secondary voltage.

   ⚠ **OBSERVATION**   Waveform:

   ac_{sec} = _____ V_{rms}.

4. Measure the dc voltage output across R_L (dc_{out}).

   ⚠ **OBSERVATION** dc_{out} = _____ V.

   ⚠ **CONCLUSION**   Current (*flows, does not flow*) _____ through R_L on both alternations of each ac input cycle. This means the bridge rectifier is a (*half-wave, full-wave*) _____ rectifier. According to theory, if there were no diode voltage drops, the average dc output of the bridge circuit without filtering should be _____ × V_{rms} of the applied ac. Does your measured value of dc_{out} for this step agree reasonably with theory? _____. How do you account for most of the difference, if any? _____

   _____

   _____.

5. Connect the oscilloscope across R_L. Sketch the waveform for two complete cycles in the "Observation" section. Indicate the maximum and minimum voltage levels.

   ⚠ **OBSERVATION**   Waveform:

▲ CONCLUSION The frequency of the waveform across the output resistor is (*one-half, equal to, double*) _____ the secondary frequency.

6. Connect the 1.0-μF filter capacitor across R_L as shown in Figure 66-2.

7. Measure and record the ac voltage at the secondary (ac_{sec}) and the dc voltage output across R_L (dc_{out}). Sketch two complete ac cycles of the oscilloscope waveform found across R_L.

▲ OBSERVATION
ac_{sec} = _____ V_{rms}.
dc_{out} = _____ V_{rms}.

Waveform:

▲ CONCLUSION What is the highest voltage output we could get from this circuit with a secondary voltage of 7 V_{rms} and no load current? About _____ V. What is the value of load current according to our measured value of V_{out}? _____ mA. The filter capacitor causes the dc level of V_{out} to be (*higher, lower*) _____ than if it were not in the circuit. This filter capacitor has (*more, less*) _____ time to discharge between charging pulses in a bridge circuit than in a half-wave rectifier circuit. The reason for this answer is that the bridge rectifier has _____ charging pulses for each complete cycle of the ac input waveform. The capacitor in a bridge rectifier will not discharge to as low a value of voltage between charging pulses as in a half-wave rectifier. When compared to a half-wave rectifier, the average voltage output of a bridge rectifier will be (*higher, lower*) _____ and remain closer to (*peak, effective, average*) _____ of the secondary voltage than for a half-wave rectifier having the same R_L and filter capacitor.

Power Supplies
Voltage Multiplier

PROJECT 67

Name: _____ Date: _____

FIGURE 67-1

PROJECT PURPOSE In this project you will observe the action of a half-wave cascade voltage doubler.

PARTS NEEDED
- ☐ DMM/VOM
- ☐ Oscilloscope
- ☐ CIS
- ☐ Function generator or audio oscillator
- ☐ Diodes: Silicon, 1-amp rating (2)
- ☐ Resistor 100 kΩ
- ☐ Capacitors 1.0 μF (2) 10 μF

PROCEDURE

1. Connect the circuit shown in Figure 67-1.

2. Set the function generator (sine-wave mode) or audio oscillator to 100 Hz and V_A to 6 V_{rms}. Connect the oscilloscope across V_A and measure the peak voltage value.

   ⚠ OBSERVATION V_A (peak) = _____ V.

   ⚠ CONCLUSION If you have used a meter to set V_A to 6 V_{rms}, the peak value as measured with the oscilloscope should be _____ times the V_{rms} value. Does your measured value for V_A (peak) match up reasonably well with that ratio? _____.

3. Connect the oscilloscope across R_L. Sketch the waveform for two complete cycles in the "Observation" section. Indicate the maximum and minimum voltage values and note the peak-to-peak ripple of the output voltage:

   Peak-to-peak ripple of V_{out} = V_{out} (max) − V_{out} (min)

363

⚠ OBSERVATION

Waveform:

V_{out} (max) = _____ V.

V_{out} (min) = _____ V.

Peak-to-peak ripple of V_{out} = _____ V.

⚠ CONCLUSION

The peak value of the output voltage should be about _____ times the peak value of the ac input voltage. Do your measurements confirm this? _____. The frequency of the ripple voltage at the output of this circuit is (*one-half, equal to, twice*) _____ the frequency of the input waveform.

4. Replace capacitor C_2 with the 10-µF capacitor. If this is an electrolytic capacitor, make sure you connect the positive terminal of the capacitor to the positive output terminal (at the cathode of diode D_2).

5. Connect the oscilloscope across R_L. Measure and record the maximum and minimum voltage values. Calculate the peak-to-peak ripple of the output voltage.

⚠ OBSERVATION

V_{out} (max) = _____ V.

V_{out} (min) = _____ V.

Peak-to-peak ripple of V_{out} = _____ V.

⚠ CONCLUSION

Increasing the value of the output capacitor (*increased, decreased*) _____ the output ripple voltage. Increasing the value of the output capacitor increased the average dc output voltage. How can you account for this? _____

_____

_____.

Story Behind the Numbers
Characteristics of a 7805 Voltage Regulator

Name: _____   Date: _____

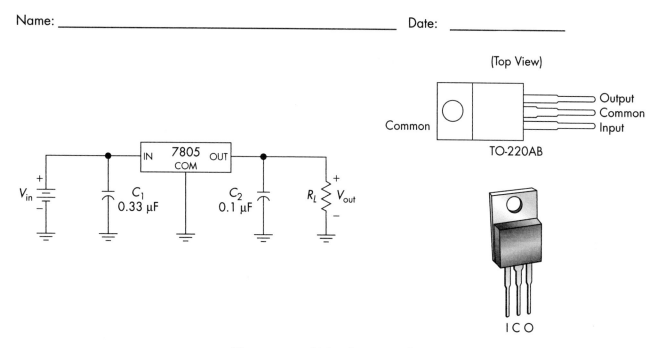

Three-terminal IC voltage regulator

Procedure

NOTE When performing the procedure steps, you have the options of using a calculator or using an Excel spreadsheet program "worksheet" for any required calculations. You may also use Excel for creating tables and for generating graphs.

1. Connect the circuit shown above. One adjustable supply (V_{in}) is used for the experiment.

2. Do not connect a load resistor to the voltage regulator output ($R_L = \infty\ \Omega$).
   A. With no load resistor, adjust V_{in} to 7 volts.
      1. Measurements. Record answers in Table 1.
         a. Measure and record the input voltage V_{in}.
         b. Measure and record the output voltage V_{out}.
      2. Calculations. Record answers in Table 1.
         a. Calculate and record the load current I_L, where $I_L = V_{out}/R_L$.
         b. Calculate and record the load power P_L, where $P_L = I_L \times V_{out}$.
         c. Calculate and record the regulator voltage drop V_{REG}, where $V_{REG} = V_{in} - V_{out}$.

d. Calculate and record the regulator power P_{REG}, where $P_{REG} = V_{REG} \times I_L + V_{in} \times I_B$. (For regulator bias current I_B, use the typical value of 4.2 mA.)

e. Calculate and record the input power P_{in}, where $P_{in} = P_{REG} + P_L$.

f. Calculate and record the voltage regulator efficiency η, where $\eta = (P_L/P_{in}) \times 100\%$.

B. With no load resistor, adjust V_{in} to 25 V.

1. Measurements. Record answers in Table 1.
   a. Measure and record the input voltage V_{in}.
   b. Measure and record the output voltage V_{out}.

2. Calculations. Record answers in Table 1.
   a. Calculate and record the load current I_L, where $I_L = V_{out}/R_L$.
   b. Calculate and record the load power P_L, where $P_L = I_L \times V_{out}$.
   c. Calculate and record the regulator voltage drop V_{REG}, where $V_{REG} = V_{in} - V_{out}$.
   d. Calculate and record the regulator power P_{REG}, where $P_{REG} = V_{REG} \times I_L + V_{in} \times I_B$. (For regulator bias current I_B, use the typical value of 4.2 mA.)
   e. Calculate and record the input power P_{in}, where $P_{in} = P_{REG} + P_L$.
   f. Calculate and record the voltage regulator efficiency η, where $\eta = (P_L/P_{in}) \times 100\%$.

3. Connect a 20-Ω (2 W) load resistor to the voltage regulator output ($R_L = 20\ \Omega$).

   A. With $R_L = 20\ \Omega$, adjust V_{in} to 7 V.

   1. Measurements. Record answers in Table 1.
      a. Measure and record the input voltage V_{in}.
      b. Measure and record the output voltage V_{out}.

   2. Calculations. Record answers in Table 1.
      a. Calculate and record the load current I_L, where $I_L = V_{out}/R_L$.
      b. Calculate and record the load power P_L, where $P_L = I_L \times V_{out}$.
      c. Calculate and record the regulator voltage drop V_{REG}, where $V_{REG} = V_{in} - V_{out}$.
      d. Calculate and record the regulator power P_{REG}, where $P_{REG} = V_{REG} \times I_L + V_{in} \times I_B$. (For regulator bias current I_B, use the typical value of 4.2 mA.)
      e. Calculate and record the input power P_{in}, where $P_{in} = P_{REG} + P_L$.
      f. Calculate and record the voltage regulator efficiency η, where $\eta = (P_L/P_{in}) \times 100\%$.

4. **Optional:** To be performed only if the power supply can provide the required current. Connect a 10-Ω (5 W) load resistor to the voltage regulator output ($R_L = 10\ \Omega$).

   A. With $R_L = 10\ \Omega$, adjust V_{in} to 7 V.

   1. Measurements. Record answers in Table 1.
      a. Measure and record the input voltage V_{in}.
      b. Measure and record the output voltage V_{out}.

   2. Calculations. Record answers in Table 1.

a. Calculate and record the load current I_L, where $I_L = V_{out}/R_L$.
b. Calculate and record the load power P_L, where $P_L = I_L \times V_{out}$.
c. Calculate and record the regulator voltage drop V_{REG}, where $V_{REG} = V_{in} - V_{out}$.
d. Calculate and record the regulator power P_{REG}, where $P_{REG} = V_{REG} \times I_L + V_{in} \times I_B$. (For regulator bias current I_B, use the typical value of 4.2 mA.)
e. Calculate and record the input power P_{in}, where $P_{in} = P_{REG} + P_L$.
f. Calculate and record the voltage regulator efficiency η, where $\eta = (P_L/P_{in}) \times 100\%$.

5. Using the data, calculate input regulation, output regulation, % load regulation, and output source resistance. Record answers in Table 2.

   Input regulation = V_{out} step 2B − V_{out} step 2A

   Output regulation = V_{out} step 2A − V_{out} step 3A

   % load regulation = (V_{out} step 2A − V_{out} step 3A) × 100%/V_{out} step 2A

   Output source resistance R_o = (V_{out} step 2A − V_{out} step 3A)/(I_L step 3A − I_L step 2A)

6. After completing the tables, answer the Analysis Questions and create the brief Technical Lab Report to complete the project.

Table 1

Measurements			Calculations					
V_{in}	V_{out}	R_L	I_L	P_L	V_{REG}	P_{REG}	P_{in}	η
V	V	Ω	A	W	V	W	W	%

Table 2

Input regulation	Output regulation	% load regulation	Output source resistance
V	V	%	Ω

Analysis Questions

NOTE Answers to these Analysis Questions should be clearly numbered and documented on separate sheets of paper with your name and the date at the top of each page. These answer sheets are to be turned in with the rest of the project documentation, as appropriate.

1. When the input voltage of the 7805 is between 2 to 20 V greater than the output voltage, the output voltage is
   a. constant
   b. 5 V
   c. regulated
   d. a, b, and c
   e. not in this list

2. According to the 7805 specifications, the minimum input voltage to provide the regulated 5-V output is _____.

3. According to the 7805 specifications, the maximum input voltage to provide the regulated 5-V output is _____.

4. For a 5-Ω load resistor connected to the 7805 with an input voltage of 7 V, calculate the
   a. load current
   b. load power
   c. regulator power
   d. input power
   e. efficiency

5. In question 4, will the voltage regulator require a heat sink? (Explain your answer.)

_____

_____

_____

Technical Lab Report

Write a brief technical lab report summarizing the technical facts learned from this project. The report should be organized to provide the following:

1. An introductory paragraph describing the type of circuit being analyzed and the key parameters that will be discussed relating to this circuit.

2. A section describing the most important characteristics of this type of circuit that were shown via the collected data in the tables and graphs.

3. Any special facts or characteristics about this type of circuit that were highlighted in answering the Analysis Questions.

4. A practical example of how the information learned in this project might help you in operating, troubleshooting, error analysis, or adjusting a circuit of this type in your home setting, in your training program setting, or in a job setting in the real world.

5. A summary statement listing the most positive aspects of the project and any parts of the project that were difficult because of equipment problems or unclear instructions. Include areas that might be improved.

Summary
Power Supplies

Name: _____ Date: _____

Complete the following review questions, indicating the appropriate response by placing a check in the box next to the correct answer.

1. The dc output voltage of a half-wave unfiltered rectifier is _____ if you neglect the forward voltage drop across the diode.
   - ☐ $0.707 \times V_{rms}$
   - ☐ $0.45 \times V_{rms}$
   - ☐ $0.318 \times V_{rms}$
   - ☐ none of these

2. The average dc output voltage of a half-wave rectifier with capacitor filtering is higher than the dc output of a half-wave rectifier without a filter.
   - ☐ True
   - ☐ False

3. A typical voltage drop across a conducting silicon diode is approximately
   - ☐ 0.0 V
   - ☐ 0.25 V
   - ☐ 0.5 V
   - ☐ 0.7 V
   - ☐ 0.9 V

4. The ripple frequency at the output of a half-wave rectifier is
   - ☐ half the ac input frequency
   - ☐ the same as the ac input frequency
   - ☐ twice the ac input frequency
   - ☐ none of these

5. The bridge rectifier is one form of
   - ☐ half-wave rectification
   - ☐ full-wave rectification
   - ☐ voltage multiplication
   - ☐ none of these

6. For one cycle of ac input, a bridge rectifier yields
   - ☐ one-half pulse of output
   - ☐ one pulse of output
   - ☐ two pulses of output
   - ☐ four pulses of output

7. Assuming a half-wave rectifier and a full-wave rectifier have identical values of output load resistance and filter capacitance, which will yield the lesser amount of output ripple voltage?
   - ☐ The half-wave rectifier
   - ☐ The full-wave rectifier

8. In a bridge rectifier circuit, how many diodes are conducting during a given alternation of the input ac waveform?
   - ☐ One
   - ☐ Two
   - ☐ Three
   - ☐ All four

9. Neglecting the forward voltage drop across the diodes, the peak output voltage of a half-wave cascade voltage doubler is _____ the peak voltage of the ac input waveform.
   ☐ close to one-half
   ☐ very nearly equal to
   ☐ about twice
   ☐ somewhat more than twice

10. In a half-wave cascade voltage doubler, the frequency of the output ripple is _____ times the frequency of the ac input waveform.
    ☐ one-half
    ☐ one
    ☐ two
    ☐ more than two

BJT CHARACTERISTICS

PART 21

Objectives

You will connect a BJT with proper biasing, measure the voltage levels, and calculate other voltages and currents. You will also determine the dc beta and alpha ratios for a BJT, and use an ohmmeter to test for shorted junctions in a BJT.

In completing these projects, you will connect circuits, make measurements, perform calculations, draw conclusions, and be able to answer questions about the following items related to BJT biasing:

- The polarity of connections for the emitter-base junction of a properly operating NPN transistor circuit
- The polarity of connections for the base-collector junction of a properly operating NPN transistor circuit
- Determining the currents in a BJT circuit, given certain voltage measurements and values of resistors
- Relative values of dc beta and alpha of a good BJT
- Relative resistance readings for good and shorted junctions in a BJT

Project/Topic Correlation Information

PROJECT	TEXT CHAPTER	SECTION	RELATED TEXT TOPIC(S)
68 BJT Biasing	26	26-2	Transistor Biasing Circuits
BJT Transistor Characteristics	25	25-3	BJT Operation
		25-4	BJT Characteristic Curves
		25-8	Information for the Technician

BJT Characteristics
BJT Biasing

PROJECT 68

Name: _____ Date: _____

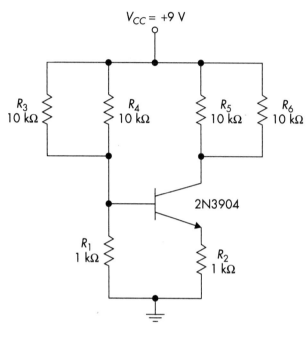

FIGURE 68-1

PROJECT PURPOSE In this project you will observe how changes in bias current of a BJT affect its operation.

PARTS NEEDED
- [ ] DMM
- [ ] VVPS (dc)
- [ ] CIS
- [ ] NPN silicon transistor: 2N3904 (or equivalent)
- [ ] Resistors
  1 kΩ (2)
  10 kΩ (4)

SPECIAL NOTE:

A normally biased BJT has the emitter-base junction forward biased and the collector-base junction reverse biased. Forward biasing, in this case, means that the N-type material is connected to the negative side of a dc source and the P-type material is connected to the positive side of the same dc source. So, with an NPN transistor, this means that the base should be positive with respect to the emitter (emitter-base bias) and the collector should be positive with respect to the base (collector-base bias).

The typical emitter-base junction voltage of a normally operating silicon transistor is about 0.7 volts. The collector-base junction voltage for the same transistor depends on the source voltage and values of the external components.

The following formulas can be helpful for completing the work in this project:

$$\text{Formula 1} \quad V_{BE} = V_B - V_E$$

where:
- V_{BE} is the base-emitter voltage
- V_B is the voltage measured from base to common
- V_E is the voltage measured from emitter to common

$$\text{Formula 2} \quad I_E = V_E/R_E$$

where:
- I_E is the emitter current
- V_E is the voltage from emitter to common
- R_E is the value of the emitter resistor

$$\text{Formula 3} \quad V_{RC} = V_{CC} - V_C$$

where:
- V_{RC} is the voltage across the collector resistor
- V_{CC} is the amount of supply voltage
- V_C is the voltage measured from the collector to common

$$\text{Formula 4} \quad I_C = V_{RC}/R_C$$

where:
- I_C is the collector current
- V_{RC} is the voltage across the collector resistor
- R_C is the value of the collector resistor

$$\text{Formula 5} \quad V_{CE} = V_C - V_E$$

where:
- V_{CE} is the collector-emitter voltage
- V_C is the voltage measured from the collector to common
- V_E is the voltage from the emitter to common

PROCEDURE

1. Connect the circuit shown in Figure 68-1.

   ⚠ CONCLUSION The total resistance between the base and $+V_{CC}$ is _____ kΩ. The total collector resistance (combination of R_5 and R_6) is _____ kΩ.

2. Set the dc source voltage to 9 V. Consider the negative side of the source as common, and measure the circuit voltages cited in the "Observation" section.

   ⚠ OBSERVATION
   V_{CC} = _____ V.   Collector to common = _____ V.
   Emitter to common = _____ V.   Base to emitter = _____ V.
   Base to common = _____ V.   Collector to emitter = _____ V.

   ⚠ CONCLUSION The base is (*positive, negative*) _____ with respect to the emitter. Is the base-to-emitter voltage close to 0.7 V? _____. Is the collector-to-emitter voltage less than V_{CC}? _____. Is the collector-to-emitter voltage greater than 0.3 V? _____. How much current must be passing through the emitter resistor? _____ mA. What is the voltage drop across the collector load resistance (V_{RC})? _____ V. This means the collector current must be approximately _____ mA. The collector is more (*positive, negative*) _____ than the base. This means the collector-base junction is (*forward, reverse*) _____ biased. The transistor is operating in (*cutoff, linear, saturation*) _____ region.

3. Remove resistor R_4 from the base circuit.

   ⚠ CONCLUSION The total resistance between the base and $+V_{CC}$ is _____ kΩ.

4. Make sure the dc source voltage is set at 9 V, and measure the circuit voltages cited in the "Observation" section.

   ⚠ OBSERVATION   Emitter to common = _____ V.   Base to emitter = _____ V.

   Base to common = _____ V.   Collector to emitter = _____ V.

   Collector to common = _____ V.

   ⚠ CONCLUSION The total resistance between the base and $+V_{CC}$ is _____ kΩ. Is the base-to-emitter voltage close to 0.7 V? _____. Is the collector-to-emitter voltage less than V_{CC}? _____. Is the collector-to-emitter voltage greater than 0.3 V? _____. How much current must be passing through the emitter resistor? _____ mA. What is the voltage drop across the collector load resistance (V_{RC})? _____ V. This means the collector current must be approximately _____ mA. Has changing the value of the base-to-V_{CC} resistance changed:

   a. The collector-to-emitter voltage? _____ If so, how? _____

   _____.

   b. The current? _____ If so, how? _____

   _____.

   c. The base-to-emitter voltage? _____. If so, how much? _____ V.

   d. The emitter current? _____. If so, how? _____

   _____.

   The transistor is operating in (*cutoff, linear, saturation*) _____ region.

5. Replace resistor R_4 as shown in Figure 68-1, and remove resistor R_6 from the collector circuit.

   ⚠ CONCLUSION The collector resistance is now _____ kΩ.

6. After making sure the dc source voltage is set at 9 V, measure the circuit voltages cited in the "Observation" section.

⚠ OBSERVATION   Emitter to common = _____ V.   Base to emitter = _____ V.

Base to common = _____ V.   Collector to emitter = _____ V.

Collector to common = _____ V.

⚠ CONCLUSION   The total resistance between the base and $+V_{CC}$ is _____ kΩ. Is the base-to-emitter voltage close to 0.7 V? _____. Is the collector-to-emitter voltage less than V_{CC}? _____. Is the collector-to-emitter voltage greater than 0.3 V? _____. How much current must be passing through the emitter resistor? _____ mA. What is the voltage drop across the collector load resistance (V_{RC})? _____ V. This means the collector current must be approximately _____ mA. Has changing the value of the collector resistance (compared with your results in Step 2) changed:

a. The collector-to-emitter voltage? _____ If so, how? _____

_____.

b. The current? _____ If so, how? _____

_____.

c. The base-to-emitter voltage? _____. If so, how much? _____ V.

d. The emitter current? _____. If so, how? _____

_____.

The transistor is operating in the (*cutoff, linear, saturation*) _____ region.

Story Behind the Numbers
BJT Transistor Characteristics

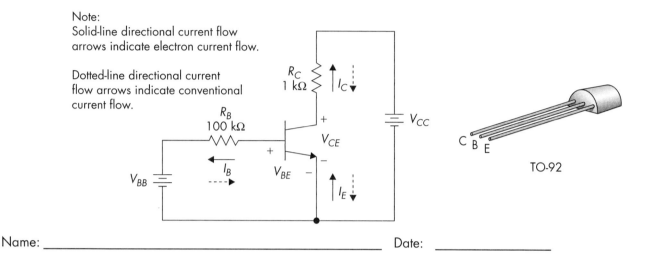

Name: _____ Date: _____

Procedure

NOTE When performing the procedure steps, you have the options of using a calculator or using an Excel spreadsheet program "worksheet" for any required calculations. You may also use Excel for creating tables and for generating graphs.

1. Connect the circuit shown above. Two separate adjustable supplies (V_{BB} and V_{CC}) are used for the experiment. It is necessary that the ground for both supplies be connected to the circuit ground.

2. Adjust V_{BB} to 1.7 V.
   A. With V_{BB} = 1.7 V, adjust V_{CC} to 0 V.
      1. Measurements. Record answers in Table 1.
         a. Measure and record the base-emitter voltage drop (V_{BE}).
         b. Measure and record the collector-emitter voltage drop (V_{CE}).
      2. Calculations. Record answers in Table 1.
         a. Calculate and record the base current I_B, where $I_B = (V_{BB} - V_{BE})/R_B$.
         b. Calculate and record the collector current I_C, where $I_C = (V_{CC} - V_{CE})/R_C$.
         c. Calculate and record the emitter current I_E, where $I_E = I_C + I_B$.
         d. Calculate the current gain β, where $\beta = I_C/I_B$.
         e. Calculate the current gain α, where $\alpha = I_C/I_E$.
   B. With V_{BB} = 1.7 V, adjust V_{CC} to 1 V.
      1. Measurements. Record answers in Table 1.
         a. Measure and record the base-emitter voltage drop (V_{BE}).
         b. Measure and record the collector-emitter voltage drop (V_{CE}).

2. Calculations. Record answers in Table 1.
   a. Calculate and record the base current I_B, where $I_B = (V_{BB} - V_{BE})/R_B$.
   b. Calculate and record the collector current I_C, where $I_C = (V_{CC} - V_{CE})/R_C$.
   c. Calculate and record the emitter current I_E, where $I_E = I_C + I_B$.
   d. Calculate the current gain β, where $\beta = I_C/I_B$.
   e. Calculate the current gain α, where $\alpha = I_C/I_E$.

C. With $V_{BB} = 1.7$ V, adjust V_{CC} to 2 V, then to 4, 6, 8, 10, 12, 14, 16, 18, and 20 V.

1. Measurements. Record answers in Table 1.
   a. Measure and record the base-emitter voltage drop (V_{BE}).
   b. Measure and record the collector-emitter voltage drop (V_{CE}).
2. Calculations. Record answers in Table 1.
   a. Calculate and record the base current I_B, where $I_B = (V_{BB} - V_{BE})/R_B$.
   b. Calculate and record the collector current I_C, where $I_C = (V_{CC} - V_{CE})/R_C$.
   c. Calculate and record the emitter current I_E, where $I_E = I_C + I_B$.
   d. Calculate the current gain β, where $\beta = I_C/I_B$.
   e. Calculate the current gain α, where $\alpha = I_C/I_E$.

Table 1

$V_{BB} = 1.7$-V data

Measurements				Calculations				
V_{BB}	V_{BE}	V_{CC}	V_{CE}	I_B	I_C	I_E	β	α
V	V	V	V	A	A	A		

3. Adjust V_{BB} to 2.7 V.
   A. With V_{BB} = 2.7 V, adjust V_{CC} to 0 V.
      1. Measurements. Record answers in Table 2.
         a. Measure and record the base-emitter voltage drop (V_{BE}).
         b. Measure and record the collector-emitter voltage drop (V_{CE}).
      2. Calculations. Record answers in Table 2.
         a. Calculate and record the base current I_B, where $I_B = (V_{BB} - V_{BE})/R_B$.
         b. Calculate and record the collector current I_C, where $I_C = (V_{CC} - V_{CE})/R_C$.
         c. Calculate and record the emitter current I_E, where $I_E = I_C + I_B$.
         d. Calculate the current gain β, where $\beta = I_C/I_B$.
         e. Calculate the current gain α, where $\alpha = I_C/I_E$.
   B. With V_{BB} = 2.7 V, adjust V_{CC} to 1 V.
      1. Measurements. Record answers in Table 2.
         a. Measure and record the base-emitter voltage drop (V_{BE}).
         b. Measure and record the collector-emitter voltage drop (V_{CE}).
      2. Calculations. Record answers in Table 2.
         a. Calculate and record the base current I_B, where $I_B = (V_{BB} - V_{BE})/R_B$.

Table 2
V_{BB} = 2.7-V data

Measurements				Calculations				
V_{BB}	V_{BE}	V_{CC}	V_{CE}	I_B	I_C	I_E	β	α
V	V	V	V	A	A	A		

b. Calculate and record the collector current I_C, where $I_C = (V_{CC} - V_{CE})/R_C$.
c. Calculate and record the emitter current I_E, where $I_E = I_C + I_B$.
d. Calculate the current gain β, where $\beta = I_C/I_B$.
e. Calculate the current gain α, where $\alpha = I_C/I_E$.

C. With $V_{BB} = 2.7$ V, adjust V_{CC} to 2 V, then to 4, 6, 8, 10, 12, 14, 16, 18, and 20 V.

1. Measurements. Record answers in Table 2.
   a. Measure and record the base-emitter voltage drop (V_{BE}).
   b. Measure and record the collector-emitter voltage drop (V_{CE}).
2. Calculations. Record answers in Table 2.
   a. Calculate and record the base current I_B, where $I_B = (V_{BB} - V_{BE})/R_B$.
   b. Calculate and record the collector current I_C, where $I_C = (V_{CC} - V_{CE})/R_C$.
   c. Calculate and record the emitter current I_E, where $I_E = I_C + I_B$.
   d. Calculate the current gain β, where $\beta = I_C/I_B$.
   e. Calculate the current gain α, where $\alpha = I_C/I_E$.

4. Adjust V_{BB} to 3.7 V.

   A. With $V_{BB} = 3.7$ V, adjust V_{CC} to 0 V.
   1. Measurements. Record answers in Table 3.
      a. Measure and record the base-emitter voltage drop (V_{BE}).
      b. Measure and record the collector-emitter voltage drop (V_{CE}).
   2. Calculations. Record answers in Table 3.
      a. Calculate and record the base current I_B, where $I_B = (V_{BB} - V_{BE})/R_B$.
      b. Calculate and record the collector current I_C, where $I_C = (V_{CC} - V_{CE})/R_C$.
      c. Calculate and record the emitter current I_E, where $I_E = I_C + I_B$.
      d. Calculate the current gain β, where $\beta = I_C/I_B$.
      e. Calculate the current gain α, where $\alpha = I_C/I_E$.

   B. With $V_{BB} = 3.7$ V, adjust V_{CC} to 1 V.
   1. Measurements. Record answers in Table 3.
      a. Measure and record the base-emitter voltage drop (V_{BE}).
      b. Measure and record the collector-emitter voltage drop (V_{CE}).
   2. Calculations. Record answers in Table 3.
      a. Calculate and record the base current I_B, where $I_B = (V_{BB} - V_{BE})/R_B$.
      b. Calculate and record the collector current I_C, where $I_C = (V_{CC} - V_{CE})/R_C$.
      c. Calculate and record the emitter current I_E, where $I_E = I_C + I_B$.
      d. Calculate the current gain β, where $\beta = I_C/I_B$.
      e. Calculate the current gain α, where $\alpha = I_C/I_E$.

   C. With $V_{BB} = 3.7$ V, adjust V_{CC} to 2 V, then to 4, 6, 8, 10, 12, 14, 16, 18, and 20 V.
   1. Measurements. Record answers in Table 3.
      a. Measure and record the base-emitter voltage drop (V_{BE}).
      b. Measure and record the collector-emitter voltage drop (V_{CE}).

2. Calculations. Record answers in Table 3.
   a. Calculate and record the base current I_B, where $I_B = (V_{BB} - V_{BE})/R_B$.
   b. Calculate and record the collector current I_C, where $I_C = (V_{CC} - V_{CE})/R_C$.
   c. Calculate and record the emitter current I_E, where $I_E = I_C + I_B$.
   d. Calculate the current gain β, where $\beta = I_C/I_B$.
   e. Calculate the current gain α, where $\alpha = I_C/I_E$.

5. Adjust V_{BB} to 4.7 V.
   A. With $V_{BB} = 4.7$ V, adjust V_{CC} to 0 V.
      1. Measurements. Record answers in Table 4 on page 385.
         a. Measure and record the base-emitter voltage drop (V_{BE}).
         b. Measure and record the collector-emitter voltage drop (V_{CE}).
      2. Calculations. Record answers in Table 4.
         a. Calculate and record the base current I_B, where $I_B = (V_{BB} - V_{BE})/R_B$.
         b. Calculate and record the collector current I_C, where $I_C = (V_{CC} - V_{CE})/R_C$.
         c. Calculate and record the emitter current I_E, where $I_E = I_C + I_B$.

Table 3
$V_{BB} = 3.7$-V data

Measurements				Calculations				
V_{BB}	V_{BE}	V_{CC}	V_{CE}	I_B	I_C	I_E	β	α
V	V	V	V	A	A	A		

d. Calculate the current gain β, where $\beta = I_C/I_B$.
e. Calculate the current gain α, where $\alpha = I_C/I_E$.

B. With $V_{BB} = 4.7$ V, adjust V_{CC} to 1 V.
1. Measurements. Record answers in Table 4.
   a. Measure and record the base-emitter voltage drop (V_{BE}).
   b. Measure and record the collector-emitter voltage drop (V_{CE}).
2. Calculations. Record answers in Table 4.
   a. Calculate and record the base current I_B, where $I_B = (V_{BB} - V_{BE})/R_B$.
   b. Calculate and record the collector current I_C, where $I_C = (V_{CC} - V_{CE})/R_C$.
   c. Calculate and record the emitter current I_E, where $I_E = I_C + I_B$.
   d. Calculate the current gain β, where $\beta = I_C/I_B$.
   e. Calculate the current gain α, where $\alpha = I_C/I_E$.

C. With $V_{BB} = 4.7$ V, adjust V_{CC} to 2 V, then to 4, 6, 8, 10, 12, 14, 16, 18, and 20 V.
1. Measurements. Record answers in Table 4.
   a. Measure and record the base-emitter voltage drop (V_{BE}).
   b. Measure and record the collector-emitter voltage drop (V_{CE}).
2. Calculations. Record answers in Table 4.
   a. Calculate and record the base current I_B, where $I_B = (V_{BB} - V_{BE})/R_B$.
   b. Calculate and record the collector current I_C, where $I_C = (V_{CC} - V_{CE})/R_C$.
   c. Calculate and record the emitter current I_E, where $I_E = I_C + I_B$.
   d. Calculate the current gain β, where $\beta = I_C/I_B$.
   e. Calculate the current gain α, where $\alpha = I_C/I_E$.

6. Using the data from Tables 1 through 4, graph the collector current (I_C) versus the collector-emitter voltage drop (V_{CE}).

7. Using the common-emitter amplifier project component values, draw its dc load line onto the BJT characteristic graph. Locate and label the Q point on the graph.

8. After completing the tables and producing the required graphs, answer the Analysis Questions and create the brief Technical Lab Report to complete the project.

Table 4
V_{BB} = 4.7-V data

Measurements				Calculations				
V_{BB}	V_{BE}	V_{CC}	V_{CE}	I_B	I_C	I_E	β	α
V	V	V	V	A	A	A		

Analysis Questions

NOTE Answers to these Analysis Questions should be clearly numbered and documented on separate sheets of paper with your name and the date at the top of each page. These answer sheets are to be turned in with the rest of the project documentation, as appropriate.

1. Evaluate the Table 1 data and answer the questions.
   a. Does V_{BE} change as V_{CE} increases?
   b. Does I_B change as V_{CE} increases?
   c. For V_{CE} > 0.3 V, does I_C change as V_{CE} increases?
   d. For V_{CE} > 0.3 V, does the current gain β change as V_{CE} increases?
   e. Is the current gain α < 1?

2. Evaluate the Table 2 data and answer the questions.
   a. Does V_{BE} change as V_{CE} increases?
   b. Does I_B change as V_{CE} increases?
   c. For V_{CE} > 0.3 V, does I_C change as V_{CE} increases?

d. For $V_{CE} > 0.3$ V, does the current gain β change as V_{CE} increases?
e. Is the current gain $\alpha < 1$?

3. Evaluate the Table 3 data and answer the questions.
   a. Does V_{BE} change as V_{CE} increases?
   b. Does I_B change as V_{CE} increases?
   c. For $V_{CE} > 0.3$ V, does I_C change as V_{CE} increases?
   d. For $V_{CE} > 0.3$ V, does the current gain β change as V_{CE} increases?
   e. Is the current gain $\alpha < 1$?

4. Evaluate the Table 4 data and answer the questions.
   a. Does V_{BE} change as V_{CE} increases?
   b. Does I_B change as V_{CE} increases?
   c. For $V_{CE} > 0.3$ V, does I_C change as V_{CE} increases?
   d. For $V_{CE} > 0.3$ V, does the current gain β change as V_{CE} increases?
   e. Is the current gain $\alpha < 1$?

5. Evaluate the data in Tables 1 through 4 and answer the questions.
   a. Does V_{BE} change as V_{BB} increases?
   b. Does I_B change as V_{BB} increases?
   c. For $V_{CE} > 0.3$ V, does I_C change as V_{BB} increases?
   d. For $V_{CE} > 0.3$ V, does the current gain β change as V_{BB} increases?

Technical Lab Report

Write a brief technical lab report summarizing the technical facts learned from this project. The report should be organized to provide the following:

1. An introductory paragraph describing the type of circuit being analyzed and the key parameters that will be discussed relating to this circuit.

2. A section describing the most important characteristics of this type of circuit that were shown via the collected data in the tables and graphs.

3. Any special facts or characteristics about this type of circuit that were highlighted in answering the Analysis Questions.

4. A practical example of how the information learned in this project might help you in operating, troubleshooting, error analysis, or adjusting a circuit of this type in your home setting, in your training program setting, or in a job setting in the real world.

5. A summary statement listing the most positive aspects of the project and any parts of the project that were difficult because of equipment problems or unclear instructions. Include areas that might be improved.

Summary
BJT Characteristics

Name: _____ Date: _____

Complete the following review questions, indicating the appropriate response by placing a check in the box next to the correct answer.

1. The emitter-base junction of an NPN transistor is said to be forward biased when
   ☐ the base is more positive than the emitter
   ☐ the emitter is more positive than the base
   ☐ the emitter and base have the same voltage

2. The base-collector junction of an NPN transistor is properly biased for normal transistor operation when
   ☐ the base is more positive than the collector
   ☐ the collector is more positive than the base
   ☐ the base and collector have the same applied voltage

3. For the NPN BJT transistor to operate in the linear region, the emitter-base and base-collector junctions must be forward biased at the same time.
   ☐ True
   ☐ False

4. When a BJT circuit operating in the linear region has a resistor of known value connected between the emitter and common, you can determine the amount of emitter current by
   ☐ measuring the emitter-to-common voltage and dividing by the amount of emitter resistance
   ☐ measuring the base-to-common voltage, measuring the emitter-to-common voltage, and dividing the difference by the amount of emitter resistance
   ☐ measuring the positive supply voltage, measuring the emitter-to-common voltage, and dividing the difference by the value of the emitter resistance

5. V_{CE} for a properly operating BJT circuit can be determined by taking the difference between the voltage measured between the collector and common and the voltage measured between the emitter and common.
   ☐ True
   ☐ False

6. The dc beta of a good BJT is
   ☐ always much greater than 1
   ☐ always a bit less than 1

7. The alpha of a good BJT is
   ☐ always much greater than 1
   ☐ always a bit less than 1

8. When the emitter-base junction of an NPN transistor is shorted, a DMM diode test will show
   ☐ 0   ☐ 0L
   ☐ 0.7

9. When the base-collector junction of an NPN transistor is in good working order, a DMM diode test will show
   ☐ 0   ☐ 0L
   ☐ 0.7

10. The DMM diode test between the emitter and collector of a good BJT will be 0L in one direction and 0.7 when the DMM leads are reversed.
    ☐ True
    ☐ False

BJT AMPLIFIER CONFIGURATIONS

PART 22

Objectives

You will connect common-emitter (CE) and common-collector (CC) BJT amplifiers and note their ac signal-amplifying characteristics.

In completing these projects, you will connect circuits, make measurements, compare input and output ac waveforms on a dual-trace oscilloscope, perform calculations, draw conclusions, and answer questions about the following items related to CE and CC amplifiers:

- The amount of phase difference between the output ac waveform and the input waveform
- The amount of gain that is characteristic of CE and CC amplifiers
- The effects that changing resistance values and supply voltage have upon the gain of the circuits
- Distinguishing a schematic diagram of a CE amplifier from that of a CC amplifier

Project/Topic Correlation Information

PROJECT	TEXT CHAPTER	SECTION	RELATED TEXT TOPIC(S)
69 Common-Emitter Amplifier	26	26-3	The Common-Emitter (CE) Amplifier
70 Common-Collector Amplifier	26	26-3	The Common-Collector (CC) Amplifier

BJT Amplifier Configurations
Common-Emitter Amplifier

PROJECT 69

Name: _____   Date: _____

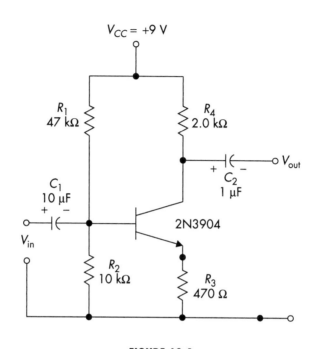

FIGURE 69-1

PROJECT PURPOSE For this project you will construct a common-emitter amplifier circuit, and you will use the DMM and dual-trace oscilloscope to determine the circuit's operating features.

PARTS NEEDED
- ☐ DMM
- ☐ VVPS (dc)
- ☐ CIS
- ☐ Dual-trace oscilloscope
- ☐ Function generator or audio oscillator
- ☐ NPN silicon transistor: 2N3904 (or equivalent)
- ☐ Resistors
  1.5 kΩ   470 Ω
  2.0 kΩ   10 kΩ
- ☐ Capacitors
  1 μF
  10 μF

PROCEDURE

DC Bias Circuit

1. Connect the circuit as shown in Figure 69-1. DO NOT connect the signal source (function generator) at this time.

392 PART 22: BJT Amplifier Configurations

2. Set the dc source for a V_{CC} of +9 V. Consider the negative side of the source as common (ground), and measure the dc voltages required in the "Observation" section.

   OBSERVATION

   V_B (voltage from base to common) = _____ V_{dc}.

   V_C (voltage from collector to common) = _____ V_{dc}.

   V_E (voltage from emitter to common) = _____ V_{dc}.

   V_{CE} (voltage from collector to emitter) = _____ V_{dc}.

   V_{BE} (voltage from base to emitter) = _____ V_{dc}.

   CONCLUSION

   Using the circuit values and the dc operating point formulas, calculate:

   $V_B =$ _____; $V_E =$ _____; $V_{CE} =$ _____; $I_E =$ _____; $r_e' =$ _____.

   Are the voltages calculated in this column close to the observed voltages (*Yes/No*)

   for: V_B _____; V_E _____; V_{CE} _____?

AC Operation

3. With the dc source still connected to V_{CC}, connect the signal source (function generator or audio oscillator) to V_{in}. Adjust the signal source for an input signal of a 0.5-V peak-to-peak, 1-kHz sine wave.

4. Connect channel 1 of the oscilloscope to V_{in}. Connect channel 2 of the oscilloscope to V_{out}. The scope ground is connected to the circuit ground. If the signal you find at V_{out} is clipped (flattened) on either or both half-cycles, **reduce the input signal level until the clipping action is no longer observed**. Sketch the waveforms and determine the readings specified in the "Observation" section.

   OBSERVATION

   V_{in} Waveform:

   V_{out} Waveform:

   $V_{in} =$ _____ V_{P-P}; $V_{out} =$ _____ V_{P-P}.

   CONCLUSION

   The oscilloscope traces indicate that the output voltage from a common-emitter amplifier is (*in phase, 180° out of phase*) _____ with its input waveform. Use the readings you found for V_{in} and V_{out} to determine the voltage gain (A_V) of this circuit. $A_V =$ _____. You have seen that the voltage gain of a common-emitter amplifier can be greater than 1. (*True, False*) _____.

   Using the values in this column and the ac operation formula, calculate the voltage gain. $A_V =$ _____.

DC Bias Circuit

5. Remove the 2.0-kΩ collector resistor (R_4) and replace it with a 1.5-kΩ resistor.

6. Remove the signal source from the circuit. Set the dc source for a V_{CC} of +9 V. Measure the dc voltages required in the "Observation" section.

⚠ OBSERVATION

$V_B = $ _____ V_{dc}. $V_{CE} = $ _____ V_{dc}.

$V_C = $ _____ V_{dc}. $V_{BE} = $ _____ V_{dc}.

$V_E = $ _____ V_{dc}.

⚠ CONCLUSION Using the circuit values and the dc operating point formulas, calculate:

$V_B = $ _____; $V_E = $ _____; $V_{CE} = $ _____; $I_E = $ _____; $r_e' = $ _____.

Use a check mark in the following list to indicate the parameter(s) that are different from step 2:

V_B _____; V_E _____; V_{CE} _____.

AC Operation

7. With the dc source still connected, connect the signal source to V_{in}. Adjust the signal source for an input signal of a 0.5-V_{P-P}, 1-kHz sine wave.

8. Connect channels 1 and 2 of the oscilloscope to V_{in} and V_{out}. If either peak of the waveform at V_{out} is clipped, reduce the input signal level until the clipping effect is no longer apparent. Gather the information requested in the "Observation" section.

⚠ OBSERVATION V_{in} Waveform:

V_{out} Waveform:

$V_{in} = $ _____ V_{P-P}; $V_{out} = $ _____ V_{P-P}.

⚠ CONCLUSION Use the readings you found for V_{in} and V_{out} to determine the voltage gain (A_V) of this circuit. $A_V = $ _____. When you increased the value of the load resistor (R_4), did you notice a significant change in the amount of voltage gain? _____. If so, describe the amount and direction of change. _____

_____

_____

_____.

Using the values in this column and the ac operation formula, calculate the voltage gain. $A_V =$ _____.

BJT Amplifier Configurations
Common-Collector Amplifier

PROJECT 70

Name: _____ Date: _____

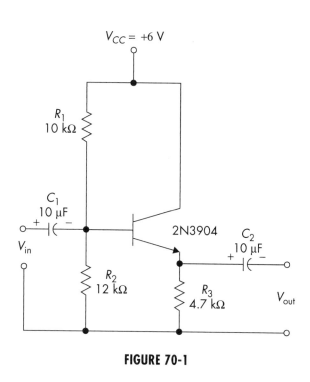

FIGURE 70-1

PROJECT PURPOSE For this project you will construct a common-collector amplifier circuit, and you will use the DMM and dual-trace oscilloscope to determine the circuit's main operating features.

PARTS NEEDED
- ☐ DMM
- ☐ VVPS (dc)
- ☐ CIS
- ☐ Dual-trace oscilloscope
- ☐ Function generator or audio oscillator
- ☐ NPN silicon transistor: 2N3904 (or equivalent)
- ☐ Resistors
  4.7 kΩ   12 kΩ
  10 kΩ
- ☐ Capacitor
  10 μF (2)

PROCEDURE

DC Bias Circuit

1. Connect the circuit as shown in Figure 70-1. DO NOT connect the signal source (function generator) at this time.

2. Set the dc source for a V_{CC} of +6 V. Consider the negative side of the source as common (ground), and measure the dc voltages required in the "Observation" section.

⚠ OBSERVATION

V_B (voltage from base to common) = _____ V_{dc}.

V_C (voltage from collector to common) = _____ V_{dc}.

V_E (voltage from emitter to common) = _____ V_{dc}.

V_{CE} (voltage from collector to emitter) = _____ V_{dc}.

V_{BE} (voltage from base to emitter) = _____ V_{dc}.

⚠ CONCLUSION

Using the circuit values and the dc operating point formulas, calculate:

V_B = _____; V_E = _____; V_{CE} = _____; I_E = _____; r_e' = _____.

Are the voltages calculated in this column close to the observed voltages (*Yes/No*) for: V_B _____; V_E _____; V_{CE} _____?

AC Operation

3. With the dc source still connected to V_{CC}, connect the signal source (function generator or audio oscillator) to V_{in}. Adjust the signal source for an input signal of a 0.5-V peak-to-peak, 1-kHz sine wave.

4. Connect channel 1 of the oscilloscope to V_{in}. Connect channel 2 of the oscilloscope to V_{out}. The scope ground is connected to the circuit ground. If the signal you find at V_{out} is clipped (flattened) on either or both half cycles, **reduce the input signal level until the clipping action is no longer observed**. Sketch the waveforms and determine the readings specified in the "Observation" section.

⚠ OBSERVATION

V_{in} Waveform:

V_{out} Waveform:

V_{in} = _____ V_{P-P}; V_{out} = _____ V_{P-P}.

⚠ CONCLUSION

The oscilloscope traces indicate that the output voltage from a common-collector amplifier is (*in phase, 180° out of phase*) _____ with its input waveform. Use the readings you found for V_{in} and V_{out} to determine the voltage gain (A_V) of this circuit. A_V = _____. (Recall that $A_V = V_{out}/V_{in}$.) You have seen in this circuit that the voltage gain of a common-collector amplifier can be greater than 1. (*True, False*) _____.

PROJECT 70: Common-Collector Amplifier

DC Bias Circuit

5. Remove the signal source from the circuit. Increase the setting of the dc source for a V_{CC} of +9 V. Measure the dc voltages required in the "Observation" section.

   ⚠ OBSERVATION $V_B = $ _____ V_{dc}. $V_{CE} = $ _____ V_{dc}.

   $V_C = $ _____ V_{dc}. $V_{BE} = $ _____ V_{dc}.

   $V_E = $ _____ V_{dc}.

   ⚠ CONCLUSION   Using the circuit values and the dc operating point formulas, calculate:

   $V_B = $ _____; $V_E = $ _____; $V_{CE} = $ _____; $I_E = $ _____; $r_e' = $ _____.

   Use a check mark in the following list to indicate the parameter(s) that are different from step 2: V_B _____; V_E _____; V_{CE} _____.

AC Operation

6. With the dc source still connected, connect the signal source to V_{in}. Adjust the signal source for an input signal of a 0.5-V_{P-P}, 1-kHz sine wave.

7. Connect channels 1 and 2 of the oscilloscope to V_{in} and V_{out}. If either peak of the waveform at V_{out} is clipped, reduce the input signal level until the clipping effect is no longer apparent. Gather the information requested in the "Observation" section.

   ⚠ OBSERVATION V_{in} Waveform:

   V_{out} Waveform:

   $V_{in} = $ _____ V_{P-P}; $V_{out} = $ _____ V_{P-P}.

   ⚠ CONCLUSION   Use the readings you found for V_{in} and V_{out} to determine the voltage gain (A_V) of this circuit. $A_V = $ _____. When you increased the value of V_{CC}, did you notice a significant change in the amount of voltage gain? _____. If so, describe the amount and direction of change; if not, describe which readings did change. _____

   _____

   _____

   _____.

Summary
BJT Amplifier Configuration

Name: _____ Date: _____

Complete the following review questions, indicating the appropriate response by placing a check in the box next to the correct answer.

1. The output ac waveform from a common-emitter amplifier is shifted _____ degrees relative to the input ac waveform.
   - ☐ 0
   - ☐ 90
   - ☐ 180
   - ☐ 360

2. The voltage gain of a common-emitter amplifier can be greater than 1.
   - ☐ True
   - ☐ False

3. Increasing the value of the collector resistor in a common-emitter amplifier _____ the voltage gain of the circuit.
   - ☐ has no effect on
   - ☐ increases
   - ☐ reduces

4. For the common-emitter circuit, the dc voltage readings show that the base-emitter junction
   - ☐ is forward biased
   - ☐ is reverse biased
   - ☐ has no bias
   - ☐ is negatively biased

5. As you decrease the amount of ac signal applied at V_{in} for a common-emitter amplifier, a corresponding decrease in V_{out} indicates that gain of the circuit is also decreasing.
   - ☐ True
   - ☐ False

6. You can identify a common-emitter amplifier by noting that
   - ☐ the emitter is connected to the negative terminal of the power source
   - ☐ the input and output terminals are both connected to the emitter
   - ☐ neither the signal input nor the signal output is connected to the emitter
   - ☐ none of these is true

7. The output ac waveform from a common-collector amplifier is shifted _____ degrees relative to the input ac waveform.
   - ☐ 0
   - ☐ 90
   - ☐ 180
   - ☐ 360

8. A common-collector amplifier is also known as a voltage follower because the output voltage "follows" the input voltage.
   - ☐ True
   - ☐ False

9. Increasing the value of V_{CC} for a common-collector amplifier
   - ☐ increases the voltage gain
   - ☐ has little, if any, effect on the voltage gain

10. You can identify an NPN common-collector amplifier by noting that
    - ☐ the collector is connected directly to the positive terminal of the power source
    - ☐ the collector is grounded, or connected to circuit common
    - ☐ the output is taken from the collector
    - ☐ none of these is true

BJT AMPLIFIER CLASSES OF OPERATION

PART 23

Objectives

You will connect each of the three main classes of BJT amplifiers and investigate their dc biasing and ac signal-amplifying characteristics.

In completing these projects, you will connect circuits, make measurements, compare input and output ac waveforms on a dual-trace oscilloscope, perform calculations, draw conclusions, and answer questions about the following items related to BJT amplifier classes of operation:

- The nature of base-emitter dc biasing for the three main classes of BJT amplifier circuits
- The way the three main classes of BJT amplifiers affect the input ac waveform
- Comparisons of amplifier efficiency and distortion

Project/Topic Correlation Information

PROJECT	TEXT CHAPTER	SECTION	RELATED TEXT TOPIC(S)
71 BJT Class A Amplifier	26	26-5	Classification by Class of Operation: Class A Amplifier Operation
		26-6	Analysis of BJT Amplifiers
72 BJT Class B Amplifier	26	26-5	Class B Amplifier Operation
73 BJT Class C Amplifier	26	26-5	Class C Amplifier Operation

BJT Amplifier Classes of Operation
BJT Class A Amplifier

PROJECT 71

Name: _____ Date: _____

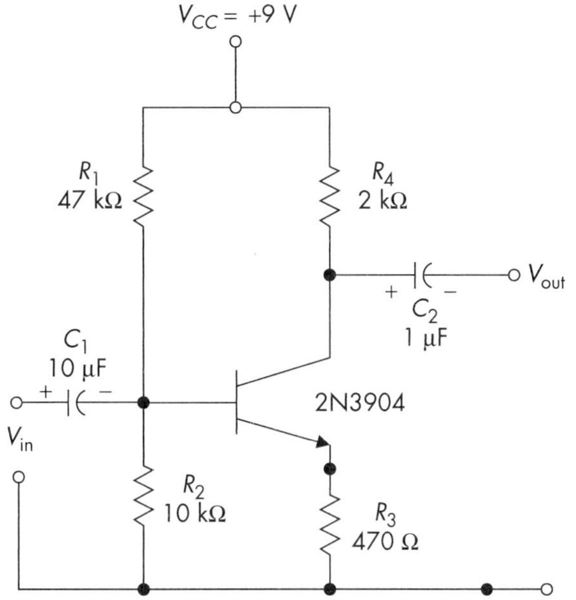

FIGURE 71-1

PROJECT PURPOSE The purpose of this project is to demonstrate the dc biasing and ac amplification of a Class A, common-emitter BJT amplifier circuit.

PARTS NEEDED
- ☐ DMM
- ☐ VVPS (dc)
- ☐ CIS
- ☐ Dual-trace oscilloscope
- ☐ Function generator or audio oscillator
- ☐ NPN silicon transistor: 2N3904 (or equivalent)
- ☐ Resistors
  - 470 Ω
  - 2 kΩ
  - 100 kΩ
  - 470 kΩ
- ☐ Capacitors
  - 1 µF
  - 10 µF

SPECIAL NOTE:

The following formula will be helpful:

$$A_V = V_{out}/V_{in}$$

where:
- A_V is the voltage gain of an amplifier
- V_{out} is the signal voltage level at the output of the amplifier (usually peak-to-peak)
- V_{in} is the signal voltage level at the input of the amplifier (usually peak-to-peak)

PROCEDURE

1. Connect the circuit as shown in Figure 71-1.

2. Set the dc source for a V_{CC} of +9 V. Make sure the signal source (function generator in the sine-wave mode or an audio oscillator) is disconnected from the circuit or set for 0-V output. Consider the negative side of the source as common, and measure the circuit's dc voltages cited in the "Observation" section.

   ⚠ **OBSERVATION**

   V_E (voltage from emitter to common) = _____ V_{dc}.

   V_C (voltage from collector to common) = _____ V_{dc}.

   V_B (voltage from base to common) = _____ V_{dc}.

   ⚠ **CONCLUSION** Use the data from "Observation" step 2 to determine: V_{CE} (voltage between collector and emitter) = _____ V_{dc}. V_{BE} (voltage between the base and emitter) = _____ V_{dc}. For Class A operation of a BJT amplifier, V_{CE} should be close to (V_{CC}, 1/2 V_{CC}, zero volts, –0.7 V) _____, so the transistor is (*conducting, not conducting*) _____ while no signal is applied at V_{in}. Confirm whether the transistor is conducting or not conducting by determining the voltage across the collector resistor (R_4) and dividing by the resistor value to determine the amount of collector current. I_C = _____ mA.

3. Connect the function generator (sine-wave mode) or audio oscillator, and use the oscilloscope to set the input signal to 1 kHz at 0.5 V peak-to-peak. If the signal you find at V_{out} is clipped on either or both alternations, **reduce the input signal level until the clipping action is no longer observed**.

4. Leave one channel of the oscilloscope connected to V_{in} and connect the second channel of the oscilloscope to V_{out}. If the signal you find at V_{out} is not a clean sinusoidal waveform, reduce the V_{in} signal level until the waveform at the collector is a clean sine wave. Sketch the waveforms and determine the readings specified in the "Observation" section.

   ⚠ **OBSERVATION** V_{in} Waveform:

V_{out} Waveform:

$V_{in} = $ _____ V_{P-P}; $V_{out} = $ _____ V_{P-P}.

▲ CONCLUSION   Comparing the phases of the waveforms at V_{in} and V_{out}, you can say they are (*in phase, 180° out of phase*) _____. What is the voltage gain of this amplifier? _____.

5. Slowly decrease the voltage level of the signal to one-half the amount used in step 4. Sketch the waveforms and determine the readings specified in the "Observation" section.

▲ OBSERVATION V_{in} Waveform:

V_{out} Waveform:

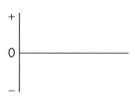

$V_{in} = $ _____ V_{P-P}; $V_{out} = $ _____ V_{P-P}.

▲ CONCLUSION   If there was distortion in the output waveform of this amplifier, could it be a sign that the input signal was too large for this circuit design. (*True, False*) _____.

BJT Amplifier Classes of Operation
BJT Class B Amplifier

Name: _____ Date: _____

FIGURE 72-1

PROJECT PURPOSE The purpose of this project is to demonstrate the dc biasing and ac amplification of a Class B, common-emitter BJT amplifier circuit.

PARTS NEEDED
- ☐ DMM
- ☐ VVPS (dc)
- ☐ CIS
- ☐ Dual-trace oscilloscope
- ☐ Function generator or audio oscillator
- ☐ NPN silicon transistor: 2N3904 (or equivalent)
- ☐ Resistors
  10 kΩ   47 kΩ
  22 kΩ
- ☐ Capacitors
  1 µF
  10 µF

PROCEDURE

1. Connect the circuit as shown in Figure 72-1.

2. Set the dc source for a V_{CC} of +9 V. Connect the signal source (function generator in the sine-wave mode or an audio oscillator) to V_{in}, but make sure its output is at zero volts. Consider the negative side of the source as common, and measure the dc voltages cited in the "Observation" section.

407

⚠ **OBSERVATION** V_E (voltage from emitter to common) = _____ V_{dc}.

V_C (voltage from collector to common) = _____ V_{dc}.

V_B (voltage from base to common) = _____ V_{dc}.

⚠ **CONCLUSION**   For Class B operation of a BJT amplifier, V_{CE} should be close to (V_{CC}, 1/2 V_{CC}, zero volts, –0.7 V) _____. This means the transistor is (*conducting, not conducting*) _____ while no signal is applied at V_{in}. Confirm whether the transistor is conducting or not conducting by determining the voltage across the collector resistor (R_2) and dividing by the resistor value to determine the amount of collector current. I_C = _____ mA.

3. Connect one channel of the oscilloscope to V_{in} and connect the second channel of the oscilloscope to V_{out}.

4. Adjust the signal source for an input signal of 1 kHz at 1.27 peak-to-peak. If the signal you find at V_{out} is clipped ("flattened") on **both** half-cycles, **reduce the input signal level until the clipping occurs on just one of the half-cycles**. Sketch the waveforms and determine the readings specified in the "Observation" section.

⚠ **OBSERVATION** V_{in} Waveform:

V_{out} Waveform:

V_{in} = _____ V_{P-P}; V_{out} = _____ V_{P-P}.

⚠ **CONCLUSION**   Comparing the phases of the waveforms at V_{in} and V_{out}, you can say they are (*in phase, 180° out of phase*) _____. The distortion noted on one alternation of the output waveform of this amplifier is a sign that something is wrong. (*True, False*) _____.

5. Connect a 47-kΩ resistor between the base of the transistor and V_{CC}. Sketch the waveform for V_{out}.

⚠ **OBSERVATION** V_{out} Waveform:

▲ CONCLUSION

Adding a resistor between the base and V_{CC} gave this amplifier some (*forward, reverse*) _____ emitter-base bias that it didn't have before. This means the amplifier is no longer operating as a Class B amplifier. (*True, False*) _____.

BJT Amplifier Classes of Operation
BJT Class C Amplifier

PROJECT 73

Name: _____ Date: _____

FIGURE 73-1

PROJECT PURPOSE The purpose of this project is to demonstrate the dc biasing and ac amplification of a Class C, common-emitter BJT amplifier circuit.

PARTS NEEDED
- ☐ DMM
- ☐ VVPS (dc)
- ☐ CIS
- ☐ Dual-trace oscilloscope
- ☐ Function generator or audio oscillator
- ☐ NPN silicon transistor: 2N3904 (or equivalent)
- ☐ 1.5-V cell
- ☐ Resistors
  22 kΩ   47 kΩ
  10 kΩ (2)
- ☐ Capacitors
  1 µF
  10 µF

PROCEDURE

1. Connect the circuit as shown in Figure 73-1.

2. Set the dc source for a V_{CC} of +9 V. Connect the signal source (function generator in its sine-wave mode or the audio oscillator) to V_{in}, but make sure its output is at zero volts. Consider the negative side of the source as common, and measure the dc voltages required in the "Observation" section.

▲ OBSERVATION V_E (voltage from emitter to common) = _____ V_{dc}.

V_C (voltage from collector to common) = _____ V_{dc}.

V_B (voltage from base to common) = _____ V_{dc}.

▲ CONCLUSION   For Class C operation of a BJT amplifier, V_{CE} for an NPN transistor should be (*equal to V_{CC}, 1/2 V_{CC}, zero volts, less than zero volts*) _____, so the transistor is (*conducting, not conducting*) _____ while no signal is applied at V_{in}. Use the formula $I_C = (V_{CC} - V_C)/R_C$ to determine the actual amount of collector current when no signal is applied. I_C = _____ mA. When there is no signal applied to this amplifier, it can be said that the emitter-base junction is (*forward, reverse*) _____ biased.

3. Connect one channel of the oscilloscope to V_{in}, and connect the second channel of the oscilloscope to V_{out}.

4. Adjust the signal source for an input signal of 1 kHz at 4.2 V peak-to-peak. If the signal you find at V_{out} is clipped on **both** alternations, **reduce the input signal level until the clipping occurs on just one of the alternations**. Sketch the waveforms and determine the readings specified in the "Observation" section.

   ▲ OBSERVATION V_{in} Waveform:

   V_{out} Waveform:

   V_{in} = _____ V_{P-P}; V_{out} = _____ V_{P-P}.

   ▲ CONCLUSION   Comparing the phases of the waveforms at V_{in} and V_{out}, you can say they are (*in phase, 180° out of phase*) _____. The distortion noted on one alternation of the output waveform of this amplifier indicates conduction during (*more than, equal to, less than*) _____ 180° of the input waveform. This type of distortion for a Class C amplifier is a sure sign that something is wrong. (*True, False*) _____.

5. Connect a 47-kΩ resistor between the base of the transistor and V_{CC}. Sketch the waveform for V_{out}.

   ⚠ OBSERVATION V_{out} Waveform:

6. Remove the 47-kΩ resistor you added in step 5 and replace it with a 10-kΩ resistor (between the base of the transistor and V_{CC}). Sketch the waveform for V_{out}.

   ⚠ OBSERVATION V_{out} Waveform:

   ⚠ CONCLUSION   As you decrease the amount of resistance between the base and V_{CC} in this circuit, the amount of (*forward, reverse*) _____ bias at the emitter-base junction becomes less, and the transistor begins conducting over a (*greater, lesser*) _____ portion of the input ac waveform.

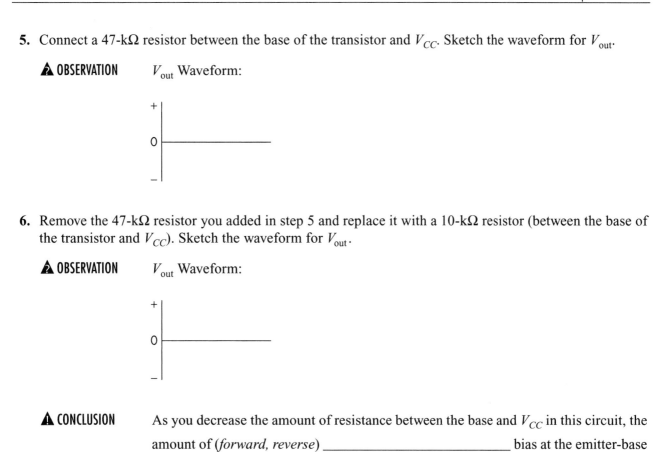

Summary
BJT Amplifier Classes of Operation

Name: _____   Date: _____

Complete the following review questions, indicating the appropriate response by placing a check in the box next to the correct answer.

1. A Class A amplifier actually amplifies _____ of an ac sine wave that is applied to the input.
   - ☐ all 360°
   - ☐ about 180°
   - ☐ less than 180°
   - ☐ none

2. The base-emitter junction of a Class A common-emitter BJT amplifier is _____ biased.
   - ☐ forward
   - ☐ reverse
   - ☐ zero

3. The output waveform of a Class A common-emitter BJT amplifier is _____ compared to its input waveform.
   - ☐ distorted
   - ☐ shifted 90°
   - ☐ shifted 180°

4. A Class A amplifier is _____ when there is no signal applied at the input.
   - ☐ conducting
   - ☐ nonconducting

5. The base-emitter junction of a Class B common-emitter BJT amplifier is _____ biased.
   - ☐ forward
   - ☐ reverse
   - ☐ zero

6. A Class B amplifier actually amplifies _____ of an ac sine wave that is applied to the input.
   - ☐ all 360°
   - ☐ about 180°
   - ☐ much less than 180°
   - ☐ none

7. A Class B amplifier is _____ when there is no signal applied at the input.
   - ☐ conducting
   - ☐ nonconducting

8. A Class C amplifier actually amplifies _____ of an ac sine wave that is applied to the input.
   - ☐ all 360°
   - ☐ about 180°
   - ☐ much less than 180°
   - ☐ none

9. The base-emitter junction of a Class C common-emitter BJT amplifier is _____ biased.
   - ☐ forward
   - ☐ reverse
   - ☐ zero

10. Of all the classes of amplifier operation, the Class C amplifier is noted for
   - ☐ the least amount of signal distortion and the highest amount of efficiency
   - ☐ the least amount of signal distortion but the least amount of efficiency
   - ☐ the greatest amount of signal distortion but the highest amount of efficiency
   - ☐ the greatest amount of signal distortion and the least amount of efficiency

JFET CHARACTERISTICS AND AMPLIFIERS

PART 24

Objectives

You will connect a JFET with proper biasing and measure the voltage levels required for plotting a pair of drain characteristic curves. You will also connect and observe the operation of a JFET amplifier circuit.

In completing these projects, you will connect circuits, make measurements, perform calculations, draw conclusions, and be able to answer questions about the following items related to JFET biasing and amplifier characteristics:

- Effects of bias-voltage levels and polarities upon the amount of drain current
- Effects of bias-voltage levels and polarities on the class of amplifier operation
- Differences between N-channel and P-channel JFET amplifier operation
- Features that distinguish classes of JFET amplifiers

Project/Topic Correlation Information

PROJECT	TEXT CHAPTER	SECTION	RELATED TEXT TOPIC(S)
74 Common-Source JFET Amplifier	27	27-1	Biasing JFET Amplifiers
			JFET Amplifier Circuits
		27-4	Power FETs
75 The Power FET JFET Characteristics	27	27-1	JFET Characteristic Curves and Ratings

JFET Characteristics and Amplifiers
Common-Source JFET Amplifier

PROJECT 74

Name: _____ Date: _____

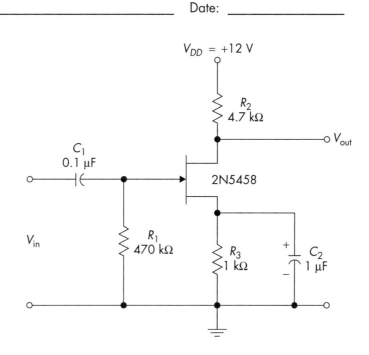

FIGURE 74-1

PROJECT PURPOSE For this project you will construct a common-source JFET amplifier circuit, and you will use the DMM and dual-trace oscilloscope to determine the circuit's main operating features.

PARTS NEEDED
- ☐ DMM
- ☐ VVPS (dc)
- ☐ Dual-trace oscilloscope
- ☐ Function generator or audio oscillator
- ☐ CIS
- ☐ N-channel JFET: 2N5458 (or equivalent)
- ☐ Resistors
  470 kΩ   4.7 kΩ
  1 kΩ
- ☐ Capacitors
  0.1 µF
  1 µF

PROCEDURE

1. Connect the circuit as shown in Figure 74-1. Set the dc source for a V_{DD} of +12 V. Connect the signal source (audio oscillator or function generator in the sine-wave mode) to V_{in}, but make sure its output is set for zero volts output.

2. Measure the dc voltages with respect to circuit common as required in the "Observation" section.

   ⚠ OBSERVATION

   V_S (voltage from source to common) = _____ V_{dc}.

   V_G (voltage from gate to common) = _____ V_{dc}.

   V_D (voltage from drain to common) = _____ V_{dc}.

   ⚠ CONCLUSION   Use the formula, $V_{DS} = V_D - V_S$, to determine the voltage drop between the source and drain. V_{DS} = _____. Based on the voltages you observed in this step, do you have good reason to suppose this is a Class A amplifier? _____. Explain your response.

   _____

   _____.

3. Connect one channel of the oscilloscope to V_{in}, and connect the second channel of the oscilloscope to V_{out}. Adjust the signal source at V_{in} for a signal of 1 kHz at 0.5 V_{P-P}. If the signal you find at V_{out} is distorted on either or both alternations, reduce the input signal level until the distortion is no longer observed. Sketch the waveforms and determine the readings specified in the "Observation" section.

   ⚠ OBSERVATION V_{in} Waveform:

   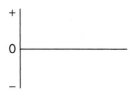

   V_{out} Waveform:

   V_{in} = _____ V_{P-P}.

   V_{out} = _____ V_{P-P}.

   ⚠ CONCLUSION   The oscilloscope traces indicate that the output voltage from a common-source amplifier is (*in phase, 180° out of phase*) _____ with its input waveform. Use the readings you found for V_{in} and V_{out} to determine the voltage gain (A_V) of this circuit. A_V = _____. (Recall that $A_V = V_{out}/V_{in}$.) You have seen in this circuit that the voltage gain of a common-source amplifier can be greater than 1. (*True, False*) _____.

4. Decrease the setting of the dc source for a V_{DD} of +9 V. If either peak of the waveform at V_{out} is clipped, reduce the input signal level until the clipping effect is no longer apparent. Gather the information requested in the "Observation" section.

⚠ OBSERVATION

V_{in} Waveform:

V_{out} Waveform:

V_{in} = _____ V_{P-P}.

V_{out} = _____ V_{P-P}.

⚠ CONCLUSION

Use the readings for V_{in} and V_{out} to determine the voltage gain of this circuit. A_V = _____. Decreasing the value of V_{DD} increases the voltage gain. (*True, False*) _____.

The Power FET
Power FETs

PROJECT 75

Name: _____ Date: _____

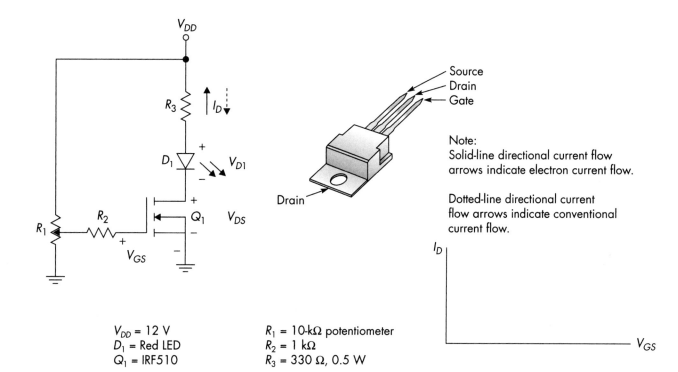

V_{DD} = 12 V
D_1 = Red LED
Q_1 = IRF510

R_1 = 10-kΩ potentiometer
R_2 = 1 kΩ
R_3 = 330 Ω, 0.5 W

Note:
Solid-line directional current flow arrows indicate electron current flow.

Dotted-line directional current flow arrows indicate conventional current flow.

FIGURE 75-1

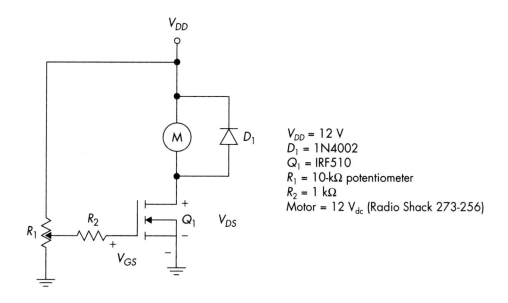

V_{DD} = 12 V
D_1 = 1N4002
Q_1 = IRF510
R_1 = 10-kΩ potentiometer
R_2 = 1 kΩ
Motor = 12 V_{dc} (Radio Shack 273-256)

FIGURE 75-2

424 PART 24: JFET Characteristics and Amplifiers

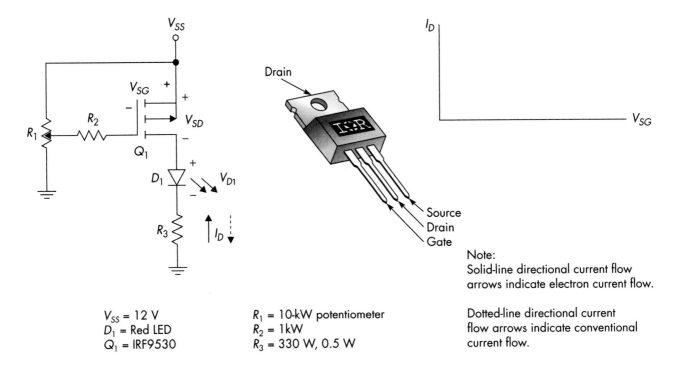

V_{SS} = 12 V
D_1 = Red LED
Q_1 = IRF9530

R_1 = 10-kΩ potentiometer
R_2 = 1 kΩ
R_3 = 330 Ω, 0.5 W

Note:
Solid-line directional current flow arrows indicate electron current flow.

Dotted-line directional current flow arrows indicate conventional current flow.

FIGURE 75-3

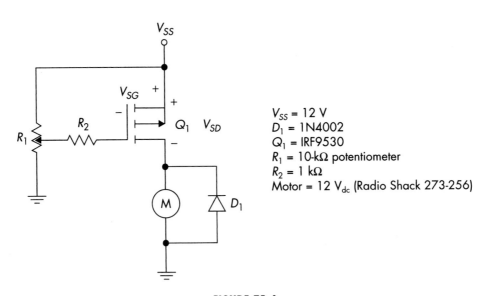

V_{SS} = 12 V
D_1 = 1N4002
Q_1 = IRF9530
R_1 = 10-kΩ potentiometer
R_2 = 1 kΩ
Motor = 12 V_{dc} (Radio Shack 273-256)

FIGURE 75-4

PROJECT PURPOSE To demonstrate the operation of the power FET to control loads.

PARTS NEEDED
- ☐ DMM
- ☐ VVPS (dc)
- ☐ Diode: Silicon 1-A (such as 1N4002)
- ☐ VVPS (dc)
- ☐ Power FET: N-channel IRF510 P-channel IRF9530N
- ☐ LED: gallium-arsenide red, 20 mA
- ☐ Resistors: 10 kΩ potentiometer 1 kΩ 330 Ω, 0.5 W
- ☐ Motor: 273-256 (Radio Shack 12 V)

PROCEDURE

1. Connect the circuit shown in Figure 75-1.

2. Apply the 12-V supply to the circuit (V_{DD}).

3. Adjust the potentiometer until the gate-source voltage $V_{GS} = 0$ V. Measure the drain-source voltage (V_{DS}), and the voltage across R_3 (V_{R_3}). Increase V_{GS} in steps indicated in the "Observation" section. Calculate the LED voltage (V_{D_1}), the drain current (I_D), $R_{DS(on)}$, and note the brightness.

⚠ OBSERVATION

$V_{GS} = 0$ V.
$V_{DS} = $ _____.
$V_{R_3} = $ _____.
$V_{D_1} = $ _____.
$I_D = $ _____.
$R_{DS(on)} = $ _____.
Brightness = _____.

$V_{GS} = 2$ V.
$V_{DS} = $ _____.
$V_{R_3} = $ _____.
$V_{D_1} = $ _____.
$I_D = $ _____.
$R_{DS(on)} = $ _____.
Brightness = _____.

$V_{GS} = 4$ V.
$V_{DS} = $ _____.
$V_{R_3} = $ _____.
$V_{D_1} = $ _____.
$I_D = $ _____.
$R_{DS(on)} = $ _____.
Brightness = _____.

$V_{GS} = 8$ V.
$V_{DS} = $ _____.
$V_{R_3} = $ _____.
$V_{D_1} = $ _____.
$I_D = $ _____.
$R_{DS(on)} = $ _____.
Brightness = _____.

$V_{GS} = 12$ V.
$V_{DS} = $ _____.
$V_{R_3} = $ _____.
$V_{D_1} = $ _____.
$I_D = $ _____.
$R_{DS(on)} = $ _____.
Brightness = _____.

⚠ CONCLUSION Complete the graph in Figure 75-1, which shows how I_D increases with V_{GS}. The brightness of the LED seems to correlate better with the amount of (V_{GS}, V_{DS}) _____ than with the amount of (V_{GS}, V_{DS}) _____.

From the component data sheets: The threshold voltage for the power FET is _____ V. The maximum drain current I_D for the power FET is _____ A. The maximum continuous LED current is _____ A. Explain why the resistor R_3 is required in the circuit. _____

_____

_____

_____.

As V_{GS} increases, $R_{DS(on)}$ (*increases, decreases, stays the same*) _____

_____.

4. Connect the circuit shown in Figure 75-2.

5. Apply the 12-V supply to the circuit (V_{DD}).

6. Adjust the potentiometer until the gate-source voltage $V_{GS} = 0$ V. Measure the drain-source voltage. Increase V_{GS} in steps indicated in the "Observation" section. Calculate the motor voltage V_M and note the motor rotation, direction, and speed.

⚠ OBSERVATION

$V_{GS} = 0$ V. Motor rotating (*Yes/No*) _____.
$V_{DS} =$ _____. Direction (*cw/ccw*) _____.
$V_M =$ _____. Speed (*slow, fast*) _____.

$V_{GS} = 2$ V. Motor rotating (*Yes/No*) _____.
$V_{DS} =$ _____. Direction (*cw/ccw*) _____.
$V_M =$ _____. Speed (*slow, fast*) _____.

$V_{GS} = 4$ V. Motor rotating (*Yes/No*) _____.
$V_{DS} =$ _____. Direction (*cw/ccw*) _____.
$V_M =$ _____. Speed (*slow, fast*) _____.

$V_{GS} = 8$ V. Motor rotating (*Yes/No*) _____.
$V_{DS} =$ _____. Direction (*cw/ccw*) _____.
$V_M =$ _____. Speed (*slow, fast*) _____.

$V_{GS} = 12$ V. Motor rotating (*Yes/No*) _____.
$V_{DS} =$ _____. Direction (*cw/ccw*) _____.
$V_M =$ _____. Speed (*slow, fast*) _____.

⚠ CONCLUSION

The motor speed seems to correlate better with the amount of (V_{GS}, V_{DS}) _____ than with the amount of (V_{GS}, V_{DS}) _____. The voltage rating for the motor is _____. The current rating for the motor is _____. Explain why no series resistor is required for the drain-source circuit. _____

_____

_____

_____.

Did the motor direction change as V_{GS} was increased? (*Yes/No*) _____. How can the direction of motor rotation be changed? _____

_____

_____

_____.

Try your solution. Did your solution work? (*Yes/No*) _____. Now, which direction is the motor turning? (*cw/ccw*) _____.

7. Connect the circuit shown in Figure 75-3.

8. Apply the 12-V supply to the circuit (V_{SS}).

9. Adjust the potentiometer until the gate-source voltage $V_{SG} = 0$ V. Measure the source-drain voltage (V_{SD}), the voltage across R_3 (V_{R_3}). Increase V_{SG} in steps indicated in the "Observation" section. Calculate the LED voltage (V_{D_1}), the drain current (I_D), $R_{DS(on)}$ and note the brightness.

⚠ OBSERVATION

$V_{SG} = 0$ V.
$V_{SD} =$ _____.
$V_{R_3} =$ _____.
$V_{D_1} =$ _____.

$I_D =$ _____.
$R_{DS(on)} =$ _____.
Brightness = _____.

$V_{SG} = 2$ V.
$V_{SD} =$ _____.
$V_{R_3} =$ _____.
$V_{D_1} =$ _____.

$I_D =$ _____.
$R_{DS(on)} =$ _____.
Brightness = _____.

$V_{SG} = 4$ V.
$V_{SD} =$ _____.
$V_{R_3} =$ _____.
$V_{D_1} =$ _____.

$I_D =$ _____.
$R_{DS(on)} =$ _____.
Brightness = _____.

$V_{SG} = 8$ V.
$V_{SD} =$ _____.
$V_{R_3} =$ _____.
$V_{D_1} =$ _____.

$I_D =$ _____.
$R_{DS(on)} =$ _____.
Brightness = _____.

$V_{SG} = 12$ V.
$V_{SD} =$ _____.
$V_{R_3} =$ _____.
$V_{D_1} =$ _____.

$I_D =$ _____.
$R_{DS(on)} =$ _____.
Brightness = _____.

428 PART 24: JFET Characteristics and Amplifiers

⚠ CONCLUSION Complete the graph in Figure 75-3, which shows how I_D increases with V_{SG}. The brightness of the LED seems to correlate better with the amount of (V_{SG}, V_{SD}) _____ than with the amount of (V_{SG}, V_{SD}) _____.

From the component data sheets: The threshold voltage for the power FET is _____ V. The maximum drain current I_D for the power FET is _____ A. The maximum continuous LED current is _____ A. Explain why the resistor R_3 is required in the circuit. _____
_____
_____
_____.

As V_{SG} increases, $R_{DS(on)}$ (*increases, decreases, stays the same*) _____
_____.

10. Connect the circuit shown in Figure 75-4.

11. Apply the 12-V supply to the circuit (V_{SS}).

12. Adjust the potentiometer until the gate-source voltage $V_{SG} = 0$ V. Measure the source-drain voltage (V_{SD}). Increase V_{SG} in steps indicated in the "Observation" section. Calculate the motor voltage V_M and note the motor rotation, direction, and speed.

⚠ OBSERVATION

$V_{SG} = 0$ V. Motor rotating (*Yes/No*) _____.
$V_{SD} =$ _____. Direction (*cw/ccw*) _____.
$V_M =$ _____. Speed (*slow, fast*) _____.

$V_{SG} = 2$ V. Motor rotating (*Yes/No*) _____.
$V_{SD} =$ _____. Direction (*cw/ccw*) _____.
$V_M =$ _____. Speed (*slow, fast*) _____.

$V_{SG} = 4$ V. Motor rotating (*Yes/No*) _____.
$V_{SD} =$ _____. Direction (*cw/ccw*) _____.
$V_M =$ _____. Speed (*slow, fast*) _____.

$V_{SG} = 8$ V. Motor rotating (*Yes/No*) _____.
$V_{SD} =$ _____. Direction (*cw/ccw*) _____.
$V_M =$ _____. Speed (*slow, fast*) _____.

$V_{SG} = 12$ V. Motor rotating (*Yes/No*) _____.
$V_{SD} =$ _____. Direction (*cw/ccw*) _____.
$V_M =$ _____. Speed (*slow, fast*) _____.

▲ CONCLUSION

The motor speed seems to correlate better with the amount of (V_{SG}, V_{SD}) _____ than with the amount of (V_{SG}, V_{SD}) _____. The voltage rating for the motor is _____. The current rating for the motor is _____.

Explain why no series resistor is required for the source-drain circuit. _____
_____
_____
_____.

Did the motor direction change as V_{SG} was increased? (*Yes/No*) _____. How can the direction of motor rotation be changed? _____
_____
_____
_____.

Try your solution. Did your solution work? (*Yes/No*) _____. Now, which direction is the motor turning? (*cw/ccw*) _____.

Story Behind the Numbers
JFET Transistor Characteristics

Name: _____   Date: _____

JFET characteristic test circuit

Procedure

NOTE When performing the procedure steps, you have the options of using a calculator or using an Excel spreadsheet program "worksheet" for any required calculations. You may also use Excel for creating tables and for generating graphs. Connections for the tri-power supply are shown in Figure 75-5 on page 437.

1. Connect the circuit shown above. Two separate adjustable supplies (V_{GG} and V_{DD}) are used for the experiment. It is necessary that the ground for both supplies be connected to the circuit ground.

2. Adjust V_{DD} to 24 V.
   A. With V_{DD} = 24 V, adjust V_{GG} to 0 V.
      1. Measurements. Record answers in Table 1.
         a. Measure and record the gate-source voltage drop (V_{GS}).
         b. Measure and record the drain-source voltage drop (V_{DS}).
      2. Calculations. Record answers in Table 1.
         a. Calculate and record the gate current I_G, where $I_G = (V_{GG} - V_{GS})/R_G$.
         b. Calculate and record the drain current I_D, where $I_D = (V_{DD} - V_{DS})/R_D$.
         c. Calculate and record the emitter current I_S, where $I_S = I_D + I_G$.
   B. With V_{DD} = 24 V, adjust V_{GG} to – 0.5 V.
      1. Measurements. Record answers in Table 1.
         a. Measure and record the gate-source voltage drop (V_{GS}).
         b. Measure and record the drain-source voltage drop (V_{DS}).

2. Calculations. Record answers in Table 1.
   a. Calculate and record the gate current I_G, where $I_G = (V_{GG} - V_{GS})/R_G$.
   b. Calculate and record the drain current I_D, where $I_D = (V_{DD} - V_{DS})/R_D$.
   c. Calculate and record the emitter current I_S, where $I_S = I_D + I_G$.

C. With $V_{DD} = 24$ V, adjust V_{GG} to -1 V, then to -1.5 V, -2 V, and so on, in 0.5 V increments until $I_D = 0$ A.
   1. Measurements. Record answers in Table 1.
      a. Measure and record the gate-source voltage drop (V_{GS}).
      b. Measure and record the drain-source voltage drop (V_{DS}).
   2. Calculations. Record answers in Table 1.
      a. Calculate and record the gate current I_G, where $I_G = (V_{GG} - V_{GS})/R_G$.
      b. Calculate and record the drain current I_D, where $I_D = (V_{DD} - V_{DS})/R_D$.
      c. Calculate and record the emitter current I_S, where $I_S = I_D + I_G$.

3. Adjust V_{GG} to 0 V.
   A. With $V_{GG} = 0$ V, adjust V_{DD} to 0 V.
      1. Measurements. Record answers in Table 2.
         a. Measure and record the gate-source voltage drop (V_{GS}).
         b. Measure and record the drain-source voltage drop (V_{DS}).
      2. Calculations. Record answers in Table 2.
         a. Calculate and record the gate current I_G, where $I_G = (V_{GG} - V_{GS})/R_G$.
         b. Calculate and record the drain current I_D, where $I_D = (V_{DD} - V_{DS})/R_D$.
         c. Calculate and record the emitter current I_S, where $I_S = I_D + I_G$.
   B. With $V_{GG} = 0$ V, adjust V_{DD} to 1 V.
      1. Measurements. Record answers in Table 2.
         a. Measure and record the gate-source voltage drop (V_{GS}).
         b. Measure and record the drain-source voltage drop (V_{DS}).
      2. Calculations. Record answers in Table 2.
         a. Calculate and record the gate current I_G, where $I_G = (V_{GG} - V_{GS})/R_G$.
         b. Calculate and record the drain current I_D, where $I_D = (V_{DD} - V_{DS})/R_D$.
         c. Calculate and record the emitter current I_S, where $I_S = I_D + I_G$.
   C. With $V_{GG} = 0$ V, adjust V_{DD} to 2 V, then to 4, 6, 8, 10, 12, 14, 16, 18, 20, 22, and 24 V.
      1. Measurements. Record answers in Table 2.
         a. Measure and record the gate-source voltage drop (V_{GS}).
         b. Measure and record the drain-source voltage drop (V_{DS}).
      2. Calculations. Record answers in Table 2.
         a. Calculate and record the gate current I_G, where $I_G = (V_{GG} - V_{GS})/R_G$.
         b. Calculate and record the drain current I_D, where $I_D = (V_{DD} - V_{DS})/R_D$.
         c. Calculate and record the emitter current I_S, where $I_S = I_D + I_G$.

4. Adjust V_{GG} to -1 V.
   A. With $V_{GG} = -1$ V, adjust V_{DD} to 0 V.
      1. Measurements. Record answers in Table 3.
         a. Measure and record the gate-source voltage drop (V_{GS}).
         b. Measure and record the drain-source voltage drop (V_{DS}).

2. Calculations. Record answers in Table 3.
   a. Calculate and record the gate current I_G, where $I_G = (V_{GG} - V_{GS})/R_G$.
   b. Calculate and record the drain current I_D, where $I_D = (V_{DD} - V_{DS})/R_D$.
   c. Calculate and record the emitter current I_S, where $I_S = I_D + I_G$.

B. With $V_{GG} = -1$ V, adjust V_{DD} to 1 V.
   1. Measurements. Record answers in Table 3.
      a. Measure and record the gate-source voltage drop (V_{GS}).
      b. Measure and record the drain-source voltage drop (V_{DS}).
   2. Calculations. Record answers in Table 3.
      a. Calculate and record the gate current I_G, where $I_G = (V_{GG} - V_{GS})/R_G$.
      b. Calculate and record the drain current I_D, where $I_D = (V_{DD} - V_{DS})/R_D$.
      c. Calculate and record the emitter current I_S, where $I_S = I_D + I_G$.

C. With $V_{GG} = -1$ V, adjust V_{DD} to 2 V, then to 4, 6, 8, 10, 12, 14, 16, 18, 20, 22, and 24 V.
   1. Measurements. Record answers in Table 3.
      a. Measure and record the gate-source voltage drop (V_{GS}).
      b. Measure and record the drain-source voltage drop (V_{DS}).
   2. Calculations. Record answers in Table 3.
      a. Calculate and record the gate current I_G, where $I_G = (V_{GG} - V_{GS})/R_G$.
      b. Calculate and record the drain current I_D, where $I_D = (V_{DD} - V_{DS})/R_D$.
      c. Calculate and record the emitter current I_S, where $I_S = I_D + I_G$.

5. Adjust V_{GG} to -2 V.
   A. With $V_{GG} = -2$ V, adjust V_{DD} to 0 V.
      1. Measurements. Record answers in Table 4.
         a. Measure and record the gate-source voltage drop (V_{GS}).
         b. Measure and record the drain-source voltage drop (V_{DS}).
      2. Calculations. Record answers in Table 4.
         a. Calculate and record the gate current I_G, where $I_G = (V_{GG} - V_{GS})/R_G$.
         b. Calculate and record the drain current I_D, where $I_D = (V_{DD} - V_{DS})/R_D$.
         c. Calculate and record the emitter current I_S, where $I_S = I_D + I_G$.
   B. With $V_{GG} = -2$ V, adjust V_{DD} to 1 V.
      1. Measurements. Record answers in Table 4.
         a. Measure and record the gate-source voltage drop (V_{GS}).
         b. Measure and record the drain-source voltage drop (V_{DS}).
      2. Calculations. Record answers in Table 4.
         a. Calculate and record the gate current I_G, where $I_G = (V_{GG} - V_{GS})/R_G$.
         b. Calculate and record the drain current I_D, where $I_D = (V_{DD} - V_{DS})/R_D$.
         c. Calculate and record the emitter current I_S, where $I_S = I_D + I_G$.
   C. With $V_{GG} = -2$ V, adjust V_{DD} to 2 V, then to 4, 6, 8, 10, 12, 14, 16, 18, 20, 22, and 24 V.
      1. Measurements. Record answers in Table 4.
         a. Measure and record the gate-source voltage drop (V_{GS}).
         b. Measure and record the drain-source voltage drop (V_{DS}).

2. Calculations. Record answers in Table 4.
   a. Calculate and record the gate current I_G, where $I_G = (V_{GG} - V_{GS})/R_G$.
   b. Calculate and record the drain current I_D, where $I_D = (V_{DD} - V_{DS})/R_D$.
   c. Calculate and record the emitter current I_S, where $I_S = I_D + I_G$.

6. Using the Table 1 data, graph the drain current (I_D) versus the gate-source voltage drop (V_{GS}).

7. Using the data from Tables 2 through 4, graph the drain current (I_D) versus the drain-source voltage drop (V_{DS}).

8. After completing the table and producing the required graphs, answer the Analysis Questions and create the brief Technical Lab Report to complete the project.

Table 1

V_{DD} = 24-V data

Measurements				Calculations		
V_{GG}	V_{GS}	V_{DD}	V_{DS}	I_G	I_D	I_S
V	V	V	V	A	A	A

Table 2
$V_{GG} = 0$-V data

Measurements				Calculations		
V_{GG}	V_{GS}	V_{DD}	V_{DS}	I_G	I_D	I_S
V	V	V	V	A	A	A

Table 3
$V_{GG} = -1$-V data

Measurements				Calculations		
V_{GG}	V_{GS}	V_{DD}	V_{DS}	I_G	I_D	I_S
V	V	V	V	A	A	A

Table 4
$V_{GG} = -2$-V data

Measurements				Calculations		
V_{GG}	V_{GS}	V_{DD}	V_{DS}	I_G	I_D	I_S
V	V	V	V	A	A	A

Analysis Questions

NOTE Answers to these Analysis Questions should be clearly numbered and documented on separate sheets of paper with your name and the date at the top of each page. These answer sheets are to be turned in with the rest of the project documentation, as appropriate.

1. The maximum amount of drain current flows through the channel region of a JFET when V_{GS} is at 0 V. (*True/False*) _____.

2. The maximum amount of drain current is called (*pinch-off current, leakage current, drain-source saturation current I_{DSS}*) _____.

3. The gate-source junction of the n-channel JFET for $V_{GG} = 0$ V (*has no bias, is reverse biased, is forward biased*) _____.

4. The gate-source junction of the n-channel JFET for $V_{GG} = -1$ V (*has no bias, is reverse biased, is forward biased*) _____.

5. An n-channel JFET is (*an enhancement-mode, a depletion mode*) _____ device.

6. For the n-channel JFET, as the gate source voltage V_{GS} increases in the negative direction, the amount of drain current I_D (*increases, decreases*) _____.

7. The gate current for the n-channel JFET is greater than 10 µA. (*True/False*) _____.

Technical Lab Report

Write a brief technical lab report summarizing the technical facts learned from this project. The report should be organized to provide the following:

1. An introductory paragraph describing the type of circuit being analyzed and the key parameters that will be discussed relating to this circuit.

2. A section describing the most important characteristics of this type of circuit that were shown via the collected data in the tables and graphs.

3. Any special facts or characteristics about this type of circuit that were highlighted in answering the Analysis Questions.

4. A practical example of how the information learned in this project might help you in operating, troubleshooting, error analysis, or adjusting a circuit of this type in your home setting, in your training program setting, or in a job setting in the real world.

5. A summary statement listing the most positive aspects of the project and any parts of the project that were difficult because of equipment problems or unclear instructions. Include areas that might be improved.

Tri-Power Supply

To produce both positive and negative voltages from a tri-power supply where each output is independent, you will connect the power supply as shown in Figure 75-5.

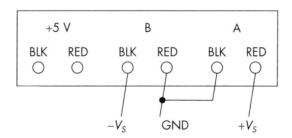

FIGURE 75-5 Tri-power supply (independent mode)

The black terminal of output A is connected to the red terminal of output B. These two terminals will be the ground for the power supply and your circuit board.

- Output A (red terminal) will be a positive voltage with respect to the circuit ground.
- Output B (black terminal) will be a negative voltage with respect to the circuit ground.

Summary
JFET Characteristics and Amplifiers

Name: _____ Date: _____

Complete the following review questions, indicating the appropriate response by placing a check in the box next to the correct answer.

1. When a JFET is properly used, the drain current is proportional to the gate current.
   - ☐ True
   - ☐ False

2. The source-gate junction of a JFET requires little bias current and signal current because it is
   - ☐ normally reverse biased
   - ☐ normally forward biased
   - ☐ normally not used
   - ☐ made of a nonconductive metal oxide film

3. A JFET should always be biased for
   - ☐ depletion-mode operation
   - ☐ enhancement-mode operation
   - ☐ sine-wave mode operation

4. For an n-channel JFET, making the gate more negative with respect to the source
   - ☐ causes drain current to increase
   - ☐ causes drain current to decrease
   - ☐ has no significant effect on drain current

5. For a common-source amplifier that uses an n-channel JFET, making the gate more negative with respect to the source causes the drain voltage to increase in the _____ direction.
   - ☐ positive
   - ☐ negative

6. The main difference between an n-channel JFET amplifier and a p-channel JFET amplifier is
   - ☐ the n-channel version inverts the incoming signal, whereas the corresponding p-channel version does not
   - ☐ the polarity of the bias voltages
   - ☐ the n-channel version operates in the depletion mode, whereas the corresponding p-channel version operates in the enhancement mode

7. A common-drain JFET amplifier
   ☐ is impractical
   ☐ has a very low input impedance
   ☐ is sometimes called a voltage follower
   ☐ inverts the input signal

8. When an n-channel JFET is operating as a Class B amplifier, you would expect to find
   ☐ a relatively large negative voltage on the gate
   ☐ a relatively small negative voltage on the gate
   ☐ a relatively large positive voltage on the gate

9. You can distinguish a common-gate JFET amplifier by noting that
   ☐ the input is applied at the gate and the output is taken from the gate
   ☐ the input is applied to the drain and the output is taken from the source
   ☐ the input is applied to the gate and the output is taken from the source
   ☐ none of these

10. Inserting a resistor in series with the source terminal of an n-channel JFET can make the source more positive than the gate. This principle is the basis for
    ☐ preventing the JFET from overheating
    ☐ increasing the current through the drain terminal
    ☐ establishing self-bias
    ☐ establishing Class C operation

OPERATIONAL AMPLIFIERS

PART 25

Objectives

You will connect basic inverting and noninverting op-amp circuits, and then you will compare their input and output signal levels and phases. You will also study the operation of a Schmitt-trigger circuit built with op-amp circuitry.

In completing these projects, you will connect circuits, observe waveforms with an oscilloscope, draw conclusions, and answer questions about the following items related to op-amp circuits:

- Explain how the location of input and feedback resistors determines whether the circuit operates as an inverting or noninverting amplifier
- Explain how the ratio of feedback to input resistance affects the voltage gain of op-amp circuits
- Describe the operation of a simple op-amp Schmitt-trigger circuit

Project/Topic Correlation Information

PROJECT	TEXT CHAPTER	SECTION	RELATED TEXT TOPIC(S)
76 Inverting Op-Amp Circuit	28	28-2	An Inverting Amplifier
77 Noninverting Op-Amp Circuit	28	28-3	A Noninverting Amplifier
78 Op-Amp Schmitt-Trigger Circuit	28	28-5	An Op-Amp Schmitt-Trigger Circuit

Operational Amplifiers
Inverting Op-Amp Circuit

PROJECT 76

Name: _____ Date: _____

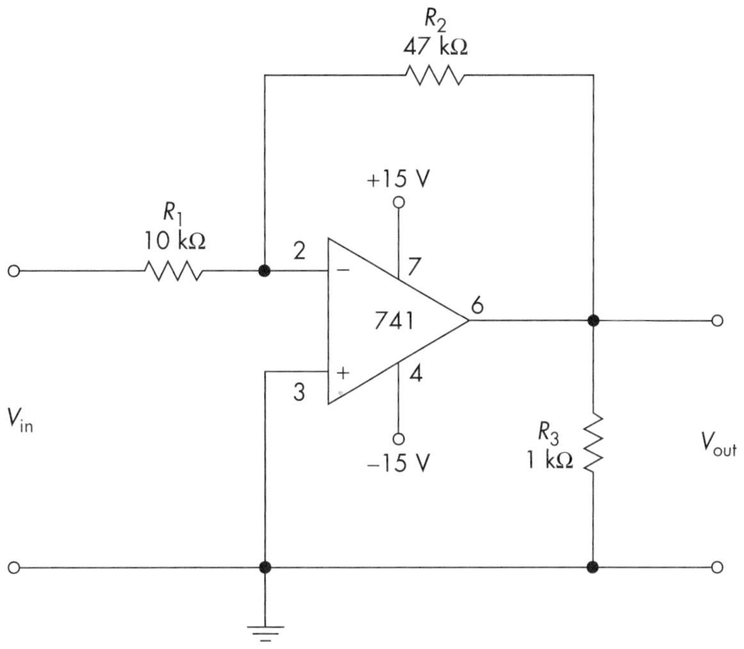

FIGURE 76-1

PROJECT PURPOSE This project demonstrates the operation of a basic inverting op-amp.

PARTS NEEDED
- [ ] DMM
- [ ] Dual-trace oscilloscope
- [ ] ±15 V DC power supply
- [ ] Function generator or audio oscillator
- [ ] CIS
- [ ] Operational amplifier: 741
- [ ] Resistors
  - 1 kΩ   47 kΩ
  - 10 kΩ   100 kΩ

SPECIAL NOTE:

You will need the following formula to complete the work:

$$A_V = -(R_2/R_1)$$

where:
- A_V is the voltage gain
- R_2 is the value of the feedback resistor
- R_1 is the value of the input resistor

444 PART 25: Operational Amplifiers

PROCEDURE

1. Connect the circuit exactly as shown in Figure 76-1. Make sure the positive terminal of the ±15-Vdc supply is connected to pin 7 of the 741 IC, the negative terminal is connected to pin 4, and the common terminal is connected to the common line as shown in the figure. Be sure you take all meter and oscilloscope readings with respect to the common line of the circuit (and not the −15-V connection of the power source). Measure and record the dc voltages requested in the "Observation" section.

   ⚠ OBSERVATION DC voltage from pin 7 to common = _____ V.

   DC voltage from pin 4 to common = _____ V.

2. Connect the function generator (sine-wave mode) or audio oscillator to V_{in}. Connect one channel of the oscilloscope to V_{in}, and connect the second channel of the oscilloscope to V_{out}. Adjust the signal source for an input signal of 1 kHz at 1 V_{P-P}. Sketch the waveforms and determine the readings specified in the Observation section.

   ⚠ OBSERVATION Peak-to-peak voltage from V_{in} to common = _____ V.

   Peak-to-peak voltage from V_{out} to common = _____ V.

   V_{in} Waveform:

   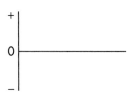

   V_{out} Waveform:

   ⚠ CONCLUSION What is the calculated voltage gain of the circuit shown in Figure 76-1? _____.
   What is the actual voltage gain of the circuit as determined by the measured values of V_{in} and V_{out}? _____. The output waveform is shifted (*0°, 90°, 180°*) _____ relative to the input waveform.

3. Increase the level of V_{in} to 2 V_{P-P}; measure and record the values of V_{in} and V_{out}.

   ⚠ OBSERVATION Peak-to-peak voltage from V_{in} to common = _____ V.

   Peak-to-peak voltage from V_{out} to common = _____ V.

   ⚠ CONCLUSION What is the actual voltage gain of the circuit as determined by the measured values of V_{in} and V_{out} in Procedure step 3? _____. Does increasing the value of V_{in} have any significant effect upon the value of V_{out}? _____. Does increasing the value of V_{in} have any significant effect upon the voltage gain of the circuit? _____.

4. Replace R_2 with a resistor having a value of 100 kΩ.

5. Adjust the signal source for 1 V_{p-p}. Sketch the waveforms and determine the readings specified in the "Observation" section.

⚠ OBSERVATION

Peak-to-peak voltage from V_{in} to common = _____ V.

Peak-to-peak voltage from V_{out} to common = _____ V.

V_{in} Waveform:

V_{out} Waveform:

⚠ CONCLUSION

What is the calculated voltage gain of the circuit when R_2 = 100 kΩ? _____.
What is the actual voltage gain of the circuit as determined by the measured values of V_{in} and V_{out}? _____. Does increasing the value of R_2 have any significant effect upon the voltage gain of the circuit? _____.

Operational Amplifiers
Noninverting Op-Amp Circuit

PROJECT 77

Name: _____ Date: _____

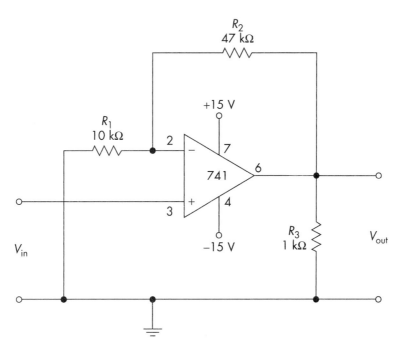

FIGURE 77-1

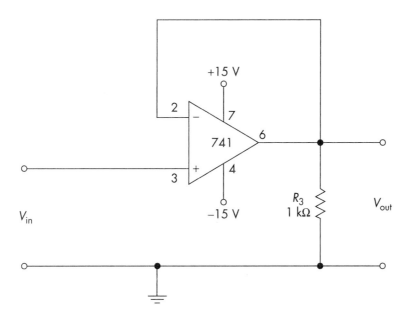

FIGURE 77-2

PROJECT PURPOSE In this project you will observe the operation of a noninverting op-amp circuit, including a voltage-follower version. You will also see the effects of overdriving an op-amp circuit that has excessive voltage gain.

PARTS NEEDED
- ☐ DMM
- ☐ Dual-trace oscilloscope
- ☐ ±15-Vdc power supply
- ☐ Function generator or audio oscillator
- ☐ CIS
- ☐ Operational amplifier: 741
- ☐ Resistors
  - 1 kΩ
  - 10 kΩ
  - 47 kΩ
  - 100 kΩ
  - 1 MΩ

SPECIAL NOTE:

The following formula can be helpful:

$$A_V = (R_2/R_1) + 1$$

where:
- A_V is the voltage gain of a noninverting op-amp
- R_2 is the value of the feedback resistor
- R_1 is the value of the input resistor

PROCEDURE

1. Connect the circuit exactly as shown in Figure 77-1. Make sure the positive terminal of the ±15-Vdc supply is connected to pin 7 of the 741 IC, the negative terminal is connected to pin 4, and the common terminal is connected to the common line as shown in the figure. Be sure you take all meter and oscilloscope readings with respect to the common line of the circuit (and not the −15-V connection of the power source). Measure and record the dc voltages requested in the "Observation" section.

   ⚠ OBSERVATION
   DC voltage from pin 7 to common = _____ V.
   DC voltage from pin 4 to common = _____ V.

2. Connect the function generator (sine-wave mode) or audio oscillator to V_{in}. Connect one channel of the oscilloscope to V_{in}, and connect the second channel of the oscilloscope to V_{out}. Adjust the signal source for an input signal of 1 kHz at 1 $V_{p\text{-}p}$. Sketch the waveforms and determine the readings specified in the "Observation" section.

   ⚠ OBSERVATION
   Peak-to-peak voltage from V_{in} to common = _____ V.
   Peak-to-peak voltage from V_{out} to common = _____ V.

   V_{in} Waveform:

V_{out} Waveform:

⚠ CONCLUSION   What is the calculated voltage gain of the circuit shown in Figure 77-1? _____.
What is the actual voltage gain of the circuit as determined by the measured values of
V_{in} and V_{out}? _____. The output waveform is shifted (*0°, 90°, 180°*) _____
relative to the input waveform.

3. Replace R_2 with a resistor having a value of 10 kΩ.

4. Leave the signal source at 1 V_{P-P}. Measure and record the data specified in the "Observation" section.

   ⚠ OBSERVATION   Peak-to-peak voltage from V_{in} to common = _____ V.
   Peak-to-peak voltage from V_{out} to common = _____ V.

   ⚠ CONCLUSION   What is the calculated voltage gain of the circuit when R_2 = 10 kΩ? _____. What
   is the actual voltage gain of the circuit as determined by the measured values of V_{in}
   and V_{out}? _____. Explain why this particular circuit might be called a voltage
   follower. _____
   _____
   _____.

5. Replace R_2 with a resistor having a value of 1 MΩ.

6. Leave the signal source at 1 V_{P-P}. Sketch the waveforms and determine the readings specified in the "Observation" section.

   ⚠ OBSERVATION   Peak-to-peak voltage from V_{in} to common = _____ V.
   Peak-to-peak voltage from V_{out} to common = _____ V.

   V_{in} Waveform:

   V_{out} Waveform:

PART 25: Operational Amplifiers

⚠ CONCLUSION What is the calculated voltage gain of the circuit when $R_2 = 1\ \text{M}\Omega$? _____.
Explain why the output waveform is distorted. _____
_____
_____.

7. Connect the circuit exactly as shown in Figure 77-2.

8. Connect the function generator to V_{in}. Set the function generator to 1 V_{P-P} and 1-kHz sine wave. Connect scope channel 1 to V_{in} and scope channel 2 to V_{out}. Sketch the V_{in} and V_{out} waveforms.

⚠ OBSERVATION V_{in} Waveform:

V_{out} Waveform:

⚠ CONCLUSION What is the calculated voltage gain of the circuit? _____. What is the actual voltage gain of the circuit as determined by the measured values of V_{in} and V_{out}? _____. Explain why this circuit might be called a unity gain analog buffer. _____
_____
_____.

Operational Amplifiers
Op-Amp Schmitt-Trigger Circuit

PROJECT 78

Name: _____   Date: _____

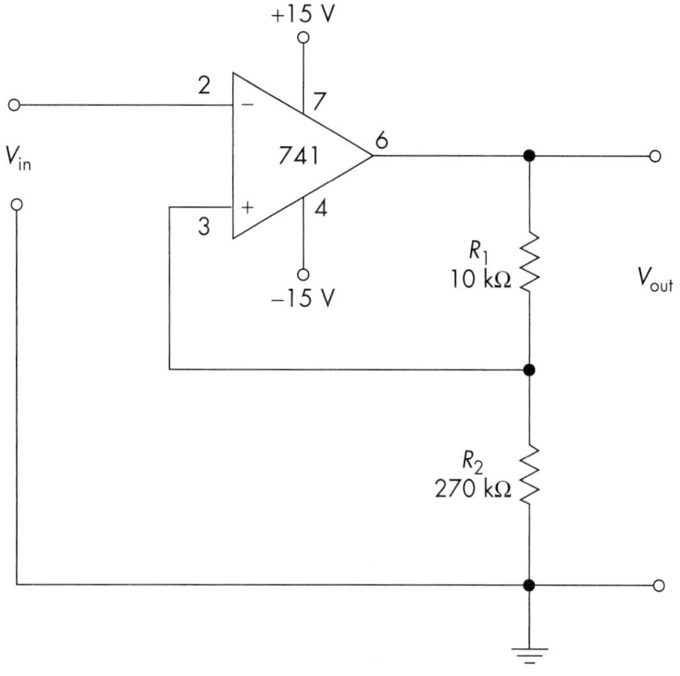

FIGURE 78-1

PROJECT PURPOSE In this project you will observe the application of an op-amp as a Schmitt-trigger circuit.

PARTS NEEDED
- ☐ DMM
- ☐ Dual-trace oscilloscope
- ☐ ±15-Vdc power supply
- ☐ Function generator or audio oscillator
- ☐ CIS
- ☐ Operational amplifier: 741
- ☐ Resistors
  10 kΩ
  270 kΩ

PROCEDURE

1. Connect the circuit exactly as shown in Figure 78-1. Make sure the positive terminal of the ±15-Vdc supply is connected to pin 7 of the 741 IC, the negative terminal is connected to pin 4, and the common terminal is connected to the common line as shown in the figure. (Be sure you take all meter and oscilloscope readings with respect to the common line of the circuit.)

2. Measure and record the dc voltages requested in the "Observation" section.

   OBSERVATION   DC voltage from pin 7 to common = _____ V.

   DC voltage from pin 4 to common = _____ V.

   CONCLUSION   Explain why having an ac input waveform makes it necessary to have both positive and negative supply voltages for an op-amp circuit. _____
   _____
   _____.

3. Connect the function generator (sine-wave mode) or audio oscillator to V_{in}. Connect one channel of the oscilloscope to V_{in} and connect the second channel of the oscilloscope to V_{out}. Adjust the signal source for an input signal of 1 kHz at 1 V_{p-p}.

4. Sketch the waveforms and determine the readings specified in the "Observation" section.

   OBSERVATION   Peak-to-peak voltage from V_{in} to common = _____ V.

   Peak-to-peak voltage from V_{out} to common = _____ V.

   V_{in} Waveform:

   V_{out} Waveform:

   CONCLUSION   Calculate the following values for the Schmitt-trigger circuit in Figure 78-1. *UTP* (calculated) = _____ V. *LTP* (calculated) = _____ V. Hysteresis = _____ V. If you superimpose oscilloscope waveforms for V_{in} and V_{out}, you should be able to determine the actual values of: *UTP* (measured) = _____ V. *LTP* (measured) = _____ V. Hysteresis = _____ V. The polarity of the output waveform is shifted (*0°, 90°, 180°*) _____ relative to the input waveform. This is an example of a nonlinear amplifier circuit. (*True, False*) _____ .

5. Adjust the signal source for an input signal 0.1 V_{p-p}. Sketch the waveforms and determine the readings specified in the "Observation" section.

   OBSERVATION   Peak-to-peak voltage from V_{in} to common = _____ V.

   Peak-to-peak voltage from V_{out} to common = _____ V.

V_{in} Waveform:

V_{out} Waveform:

⚠ CONCLUSION Describe how the waveforms are different from those in Procedure step 4. _____
_____
_____.

How do you account for the differences? _____
_____
_____.

Summary
Operational Amplifiers

Name: _____ Date: _____

Complete the following review questions, indicating the appropriate response by placing a check in the box next to the correct answer.

1. For an inverting op-amp
   ☐ the input and feedback resistors are both connected to the noninverting input
   ☐ the input resistor is connected to the inverting input and the feedback resistor is connected to the noninverting input
   ☐ the input resistor is connected to the noninverting input and the feedback resistor is connected to the inverting input
   ☐ the input and feedback resistors are both connected to the inverting input

2. The voltage gain of an inverting op amp is determined by
   ☐ dividing the amount of feedback resistance by the amount of input resistance
   ☐ dividing the amount of input resistance by the amount of feedback resistance
   ☐ dividing the amount of input voltage by the value of the input resistance
   ☐ multiplying the amount of input voltage by the amount of output voltage

3. An inverting op-amp circuit can never be used as a voltage follower because
   ☐ the output voltage is always larger than the input voltage
   ☐ the output waveform is always out of phase with the input waveform
   ☐ the output is nonlinear

4. What is the voltage gain of an inverting op-amp circuit when the input resistance equals the feedback resistance?
   ☐ 0
   ☐ +1
   ☐ −1
   ☐ Cannot be determined without knowing the resistor values

5. For a noninverting op amp
   ☐ the input is applied to the noninverting input and the feedback resistor is connected to the noninverting input
   ☐ the input is applied to the inverting input and the feedback resistor is connected to the noninverting input
   ☐ the input is applied to the noninverting input and the feedback resistor is connected to the inverting input
   ☐ the input is applied to the inverting input and the feedback resistor is connected to the inverting input

455

6. The voltage gain of a noninverting op amp
   - ☐ can be determined by dividing the amount of feedback resistance by the amount of input resistance
   - ☐ can be determined by dividing the amount of output voltage by the amount of input voltage
   - ☐ can be determined by dividing the amount of input voltage by the value of the feedback resistance
   - ☐ is always 1

7. What is the voltage gain of a noninverting amplifier when the feedback resistance is zero ohms?
   - ☐ 0
   - ☐ 1
   - ☐ 2
   - ☐ Cannot be determined

8. As long as the output of an op-amp remains undistorted, the amount of input voltage has little to do with the amount of voltage gain for the circuit.
   - ☐ True
   - ☐ False

9. One of the main purposes of a Schmitt-trigger circuit is to transform a sine waveform into a rectangular waveform.
   - ☐ True
   - ☐ False

10. The fact that the input signal for an op-amp Schmitt-trigger circuit goes to the inverting input accounts for
    - ☐ the flattening of the output waveform
    - ☐ the very high voltage gain of the circuit
    - ☐ the 180° phase shift of the waveform
    - ☐ none of these

OSCILLATORS AND MULTIVIBRATORS

PART 26

Objectives

In completing these projects, you will connect circuits, make oscilloscope measurements, perform calculations, draw conclusions, and answer questions about the following items related to Wien-bridge oscillators and 555-type astable multivibrators:

- Point out which components in the circuits determine their operating frequencies
- Describe how each frequency-determining component affects the operating frequency
- Explain the basic principles of lead-lag phase shift in a Wien-bridge oscillator
- Explain the basic principles of capacitor charge and discharge in a 555-type astable multivibrator

Project/Topic Correlation Information

PROJECT	TEXT CHAPTER	SECTION	RELATED TEXT TOPIC(S)
79 **Wien-Bridge Oscillator Circuit**	29	29-4	Wien-Bridge Oscillator
80 **555 Astable Multivibrator**	29	29-5	Astable Multivibrators

Oscillators and Multivibrators
Wien-Bridge Oscillator Circuit

PROJECT 79

Name: _____ Date: _____

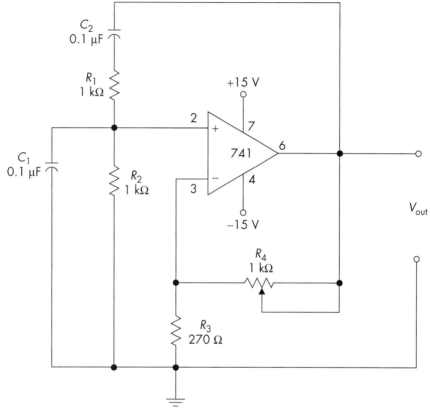

FIGURE 79-1

PROJECT PURPOSE This project gives you a chance to check out the operation of a Wien-bridge oscillator circuit.

PARTS NEEDED
- ☐ Oscilloscope
- ☐ ±15-Vdc power supply
- ☐ CIS
- ☐ Operational amplifier: 741
- ☐ 1-kΩ linear potentiometer
- ☐ Capacitors
  0.1 µF (2)
- ☐ Resistors
  270 Ω
  1 kΩ (2)
  2.7 kΩ (2)
  27 kΩ

460 PART 26: Oscillators and Multivibrators

SPECIAL NOTE:

You will need the following formula for calculating the operating frequency of a Wien-bridge oscillator:

$$f_r = 1/2\pi RC$$

where:
- f_r is the operating frequency of the circuit
- R is the value assigned to both resistors in the lead-lag network
- C is the value assigned to both capacitors in the lead-lag network

PROCEDURE

1. Connect the circuit shown in Figure 79-1.

   ⚠ CONCLUSION   List the components that make up the lead-lag network. _____
   _____.

   This op amp is connected as (*an inverting, a noninverting*) _____ amplifier.

2. Connect the oscilloscope to V_{out} and adjust potentiometer R_3 until you see a maximum level of stable oscillation.

3. Sketch two complete cycles of the output waveform in the space provided in the Observation section.

   ⚠ OBSERVATION   Output Waveform:

   ```
   +|
    |
   0|————————
    |
   -|
   ```

   ⚠ CONCLUSION   Describe the form of the output waveform as sinusoidal, rectangular, sawtooth, or triangular: _____.

4. Measure the peak-to-peak amplitude of the waveform at V_{out}.

   ⚠ OBSERVATION V_{out} (peak-to-peak) = _____ V.

   ⚠ CONCLUSION   The amplitude of the output waveform is mainly determined by (*the gain of the amplifier, the values of the components in the lead-lag circuit*) _____
   _____.

5. Measure the period of the waveform at V_{out}.

   ⚠ OBSERVATION   *Period* = _____ ms.

   ⚠ CONCLUSION   Calculate the operating frequency of this circuit according to the formula for the operating frequency of a Wien-bridge oscillator. _____ kHz. Use the formula,

$f = 1/period$, to determine the actual operating frequency. _____ kHz. Account for the difference, if any, between the calculated and measured frequency. _____
_____.

6. Replace resistors R_1 and R_2 in the circuit with 2.7-kΩ resistors. Adjust potentiometer R_3, if necessary, to get a maximum level of stable oscillation.

   ⚠ CONCLUSION   According to the formula for the operating frequency of a Wien-bridge oscillator, increasing the value of the lead-lag resistors should (*increase, decrease, have no effect on*) _____ the operating frequency.

7. Measure the period of the waveform at the output.

   ⚠ OBSERVATION   *Period* = _____ ms.

   ⚠ CONCLUSION   The actual operating frequency of the circuit is now _____ Hz.

Oscillators and Multivibrators
555 Astable Multivibrator

PROJECT 80

Name: _____   Date: _____

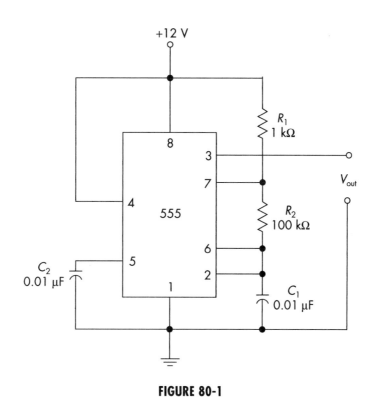

FIGURE 80-1

PROJECT PURPOSE In this project you will use a 555 timer IC to construct and observe waveforms for an astable multivibrator.

PARTS NEEDED
- ☐ DMM
- ☐ Oscilloscope
- ☐ VVPS (dc)
- ☐ CIS
- ☐ 555 timer IC
- ☐ Resistors
  1 kΩ   100 kΩ
  10 kΩ
- ☐ Capacitors
  0.01 µF (2)

SPECIAL NOTE:

You will need the following formula for calculating the operating frequency of this astable multivibrator:

$$f = 1/0.69C(R_A + 2R_B)$$

where:
- f is the output frequency
- C is the value of the timing capacitor
- R_A and R_B are the values of the timing resistors

463

464 PART 26: Oscillators and Multivibrators

PROCEDURE

1. Connect the circuit exactly as shown in Figure 80-1.

   ⚠ CONCLUSION   Which resistor in the formula for operating frequency corresponds to resistor R_1 in this circuit? _____. Which corresponds to resistor R_2? _____.

2. Connect the oscilloscope to V_{out} and sketch two complete cycles of the output waveform in the space provided in the "Observation" section.

   ⚠ OBSERVATION   Output Waveform:

   +|
   0|————————
   −|

   ⚠ CONCLUSION   Describe the form of the output waveform as sinusoidal, rectangular, sawtooth, or triangular: _____.

3. Measure the peak-to-peak amplitude and period of the waveform at V_{out}.

   ⚠ OBSERVATION V_{out} (peak-to-peak) = _____ V.

   Period = _____ ms.

   ⚠ CONCLUSION   Calculate the operating frequency of this circuit according to the formula for the operating frequency of 555-type astable multivibrators. _____ kHz. Use the formula, $f = 1/\text{period}$, to determine the actual operating frequency. _____ kHz. Account for the difference, if any, between the calculated and measured frequency.

   _____

   _____.

4. Replace resistor R_2 with a 10-kΩ resistor.

   ⚠ CONCLUSION   According to the formula for the operating frequency of a 555-type astable multivibrator, decreasing the value of either resistor should (*increase, decrease, have no effect on*) _____ the operating frequency.

5. Measure the peak-to-peak amplitude and period of the waveform at V_{out}.

   ⚠ OBSERVATION V_{out} (peak-to-peak) = _____ V.

   Period = _____ ms.

   ⚠ CONCLUSION   The actual operating frequency of the circuit is now _____ Hz.

PART 26

Summary
Oscillators and Multivibrators

Name: _____ Date: _____

Complete the following review questions, indicating the appropriate response by placing a check in the box next to the correct answer.

1. At the frequency of oscillation, the lead-lag *RC* network in a Wien-bridge oscillator produces an overall phase shift of _____ degrees.
   - ☐ 0
   - ☐ 180
   - ☐ 90
   - ☐ 270

2. In an op-amp version of a Wien-bridge oscillator, the lead-lag network is connected to the _____ input of the op amp.
   - ☐ inverting
   - ☐ voltage offset
   - ☐ noninverting

3. In a Wien-bridge oscillator, decreasing the value of either or both resistors in the lead-lag network
   - ☐ increases the operating frequency
   - ☐ decreases the operating frequency
   - ☐ has no effect on the operating frequency

4. In a Wien-bridge oscillator, increasing the value of either or both capacitors in the lead-lag network
   - ☐ increases the operating frequency
   - ☐ decreases the operating frequency
   - ☐ has no effect on the operating frequency

5. In a 555 astable multivibrator, the charge path for the timing capacitor is through
   - ☐ just one timing resistor
   - ☐ both timing resistors
   - ☐ neither timing resistor

6. In a 555 astable multivibrator, the discharge path for the timing capacitor is through
   - ☐ just one timing resistor
   - ☐ both timing resistors
   - ☐ neither timing resistor

7. In a 555 astable multivibrator, decreasing the value of either or both of the timing resistors
   - ☐ increases the operating frequency
   - ☐ decreases the operating frequency
   - ☐ has no effect on the operating frequency

465

8. For a 555-type astable multivibrator, the discharge time (time the waveform is near zero volts) is always greater than the charge time (time the waveform is near the value of the positive supply voltage).
   ☐ True
   ☐ False

9. In a 555 astable multivibrator, increasing the value of the timing capacitor
   ☐ increases the operating frequency
   ☐ decreases the operating frequency
   ☐ has no effect on the operating frequency

10. According to the formulas for the operating frequency of Wien-bridge oscillators and 555-type astable multivibrators, increasing the amount of supply voltage should
    ☐ increase the operating frequency
    ☐ decrease the operating frequency
    ☐ have no effect on the operating frequency

SCR OPERATION

Objectives

You will connect basic SCR circuits and note how they can be used to control dc and ac power.

In completing these projects, you will connect and operate circuits, measure voltage levels, observe waveforms with an oscilloscope, draw conclusions, and answer questions about the following items related to SCRs:

- Conditions for starting, sustaining, and stopping the conduction of an SCR in dc and ac circuits
- Distribution of voltages and currents in SCR control circuits
- Reasons for differing levels of load power in SCR control circuits

Project/Topic Correlation Information

PROJECT	TEXT CHAPTER	SECTION	RELATED TEXT TOPIC(S)
81 SCRs in DC Circuits	30	30-1	Theory of SCR Operation SCR Circuits
82 SCRs in AC Circuits	30	30-1	SCR Circuits

SCR Operation
SCRs in DC Circuits

PROJECT 81

Name: _____ Date: _____

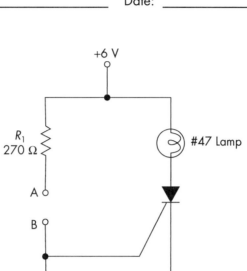

FIGURE 81-1

PROJECT PURPOSE The purpose of this project is to demonstrate how an SCR in a dc circuit cannot conduct until it is gated on and, once gated on, remains on until the power source is interrupted.

PARTS NEEDED
- ☐ DMM
- ☐ Oscilloscope
- ☐ VVPS (dc)
- ☐ CIS
- ☐ 6-V incandescent lamp: #47 (or equivalent)
- ☐ SCR: 2N5060 (or equivalent)
- ☐ Resistors
  270 Ω
  1 kΩ

SPECIAL NOTE:

Make all meter readings with respect to the common connections for the circuit.

PROCEDURE

1. Connect the circuit exactly as shown in Figure 81-1. Make sure the 6-Vdc power is turned **off**. Make sure there is **no connection** between points A and B of the circuit.

2. Turn on the dc power supply and make sure it is set for 6 V.

    NOTE ▶ The lamp should not be on at this time.

    Measure the voltages at the anode and at the gate of the SCR. Record the data in the "Observation" section.

    ⚠ OBSERVATION   Anode voltage = _____ V.

    Gate voltage = _____ V.

    The lamp is (*on, off*) _____.

    ⚠ CONCLUSION   The cathode-anode terminals of the SCR in this circuit are (*forward, reverse, not*) _____ biased. The cathode-gate terminals are (*forward, reverse, not*) _____ biased. Calculate the voltage across the lamp by subtracting the anode voltage (recorded in the "Observation" section) from the dc supply-voltage: Lamp (load) voltage = _____ V. Under the conditions observed here, the SCR must be in its (*on, off*) _____ state. This means there is (*current, no current*) _____ flowing through the SCR and through the load.

3. With 6 Vdc still applied to the circuit, make a jumper-wire connection between points A and B. Provide the data required in the "Observation" section.

    ⚠ OBSERVATION   The lamp is (*on, off*) _____.

    Anode voltage = _____ V.

    Gate voltage = _____ V.

    ⚠ CONCLUSION   The cathode-anode terminals of the SCR in this circuit are (*forward, reverse, not*) _____ biased. The cathode-gate terminals are (*forward, reverse, not*) _____ biased. Calculate the voltage across the lamp: Lamp (load) voltage = _____ V. Under the conditions observed here, the SCR is in its (*on, off*) _____ state. This means there is (*current, no current*) _____ flowing through the SCR and through the load.

4. With 6 Vdc still applied to the circuit, remove your jumper-wire connection between points A and B. Provide the data required in the "Observation" section.

    ⚠ OBSERVATION   The lamp is (*on, off*) _____.

    Anode voltage = _____ V.

    Gate voltage = _____ V.

⚠ CONCLUSION   The cathode-anode terminals of the SCR in this circuit are (*forward, reverse, not*) _____ biased. The cathode-gate terminals are (*forward, reverse, not*) _____ biased.

5. Turn off the dc power supply. Monitor the dc power supply output voltage with your voltmeter and wait for the voltage to drop all the way to zero.

6. Make sure there is **no connection** between points A and B of the circuit, and turn on the dc power supply. Provide the data required in the "Observation" section.

⚠ OBSERVATION   The lamp is (*on, off*) _____.

⚠ CONCLUSION   Explain what is necessary to get the SCR conducting once you have restored dc power to the circuit. _____
_____
_____.

SCR Operation
SCRs in AC Circuits

Name: _____ Date: _____

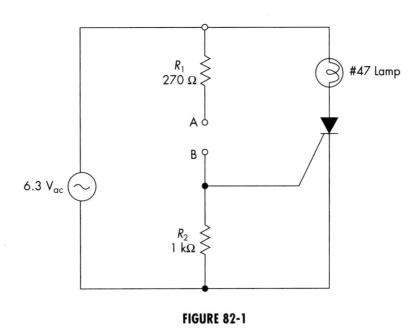

FIGURE 82-1

PROJECT PURPOSE The purpose of this project is to demonstrate that an SCR in an ac circuit cannot conduct until it is forward biased between the anode and cathode and is gated on. You will also see that the SCR is switched off when the applied ac voltage reverses polarity.

PARTS NEEDED
- ☐ DMM
- ☐ Dual-trace oscilloscope
- ☐ SCR: 2N5060 (or equivalent)
- ☐ VVPS (dc)
- ☐ 6.3-V_{ac} source
- ☐ CIS
- ☐ 6-V incandescent lamp: #47 (or equivalent)
- ☐ Resistors 270 Ω   1 kΩ

SPECIAL NOTE:

Make all oscilloscope readings with respect to the common connections for the circuit.

PROCEDURE

1. Connect the circuit exactly as shown in Figure 82-1. Connect one channel of the oscilloscope to the 6.3-V_{ac} source, and connect the second channel of the oscilloscope to the anode of the SCR. Make sure there is **no connection** between points A and B of the circuit.

2. With ac power applied to the circuit, note the state of the lamp and the waveforms. Record your findings in the "Observation" section.

   OBSERVATION   The lamp is (*on, off*) _____.

   V_{ac} Source Waveform:

   SCR Anode Waveform:

   V_{ac} voltage = _____ V_{P-P}.

   SCR anode voltage = _____ V_{P-P}.

   CONCLUSION   The SCR is evidently (*conducting on both half-cycles, conducting on every other half-cycle, not conducting at all*) _____.

   Explain your answer. _____

   _____

   _____.

   In this particular operating state, the SCR is acting as (*an open, a closed*) _____

   _____ switch.

3. Connect a jumper wire between points A and B in the circuit. With ac power still applied to the circuit, note the state of the lamp and the waveforms appearing on the oscilloscope. Record your findings in the "Observation" section.

   OBSERVATION   The lamp is (*on, off*) _____.

   V_{ac} Source Waveform:

SCR Anode Waveform:

V_{ac} voltage = _____ V_{P-P}.

SCR anode voltage = _____ V_{P-P}.

▲ CONCLUSION The SCR is evidently (*conducting on both half-cycles, conducting on every other half-cycle, not conducting at all*) _____.

Explain your answer. _____
_____
_____.

In this particular operating state, the SCR is acting as a (*full-wave, half-wave*) _____ rectifier.

4. Without interrupting the source of ac power, remove the jumper wire you connected between points A and B in the circuit. Note the circuit's response and record your findings in the "Observation" section.

▲ OBSERVATION The lamp is (*on, off*) _____.

V_{ac} Source Waveform:

SCR Anode Waveform:

V_{ac} voltage = _____ V_{P-P}.

SCR anode voltage = _____ V_{P-P}.

▲ CONCLUSION The SCR is evidently (*conducting on both half-cycles, conducting on every other half-cycle, not conducting at all*) _____.

Explain your answer. _____
_____
_____.

PART 27

Summary
SCR Operation

Name: _____ Date: _____

Complete the following review questions, indicating the appropriate response by placing a check in the box next to the correct answer.

1. In order to begin conduction, an SCR
   - ☐ must be forward biased at the anode, but the gate bias makes no difference
   - ☐ must be forward biased at the anode and the gate at the same time
   - ☐ must be forward biased at the gate, but the anode-cathode bias makes no difference

2. In order to sustain conduction of an SCR,
   - ☐ it must remain forward biased at the anode, but the gate bias makes no difference
   - ☐ it must remain forward biased at both the anode and the gate
   - ☐ it must be forward biased at the gate; however, the anode-cathode bias makes no difference

3. In order to stop conduction of an SCR,
   - ☐ the gate must be at zero volts or reverse biased; the anode bias makes no difference
   - ☐ the anode must be at zero volts or reverse biased; the gate bias makes no difference
   - ☐ the gate and the anode must be reverse biased at the same time

4. In an ac circuit, conduction of the SCR automatically stops when the ac waveform reverses the anode voltage from positive to negative.
   - ☐ True
   - ☐ False

5. For an SCR operating in a dc circuit,
   - ☐ the load current equals the source current minus the SCR current; and the load voltage equals the voltage drop across the SCR
   - ☐ the load current equals the SCR current; and the load voltage equals the source voltage minus the SCR voltage
   - ☐ the load current equals about one-half the SCR current; and the load voltage equals about one-half the source voltage

6. For an SCR that is switched on and forward biased in an ac circuit, the load voltage is equal to the instantaneous values of the source voltage minus the forward voltage drop across the SCR.
   - ☐ True
   - ☐ False

7. If an SCR operating in an ac circuit could be gated on 90° into its forward-biasing half-cycle (instead of 0°),
   ☐ the load would have more power applied to it
   ☐ the load would have less power applied to it
   ☐ there would be no effect on load power

8. At best, an SCR operating from an ac source can only conduct on alternate half-cycles (180° of the full ac waveform).
   ☐ True
   ☐ False

9. An SCR that controls an incandescent lamp in a dc circuit allows the lamp to burn at full brightness because
   ☐ nearly all dc power is being applied to the lamp
   ☐ the SCR is acting as a closed switch
   ☐ there is only a small forward voltage drop across the SCR
   ☐ all of the above are true

10. An SCR that controls an incandescent lamp in an ac circuit does not allow the lamp to burn at full brightness because
    ☐ dc power is more energetic than ac power
    ☐ the SCR cannot allow power to be applied for more than 180° of the ac waveform
    ☐ the forward voltage drop of an SCR is relatively large
    ☐ all of the above are true

FIBER-OPTIC SYSTEM CHARACTERISTICS

Story Behind the Numbers

Name: _____ Date: _____

Fiber-Optic System Characteristics

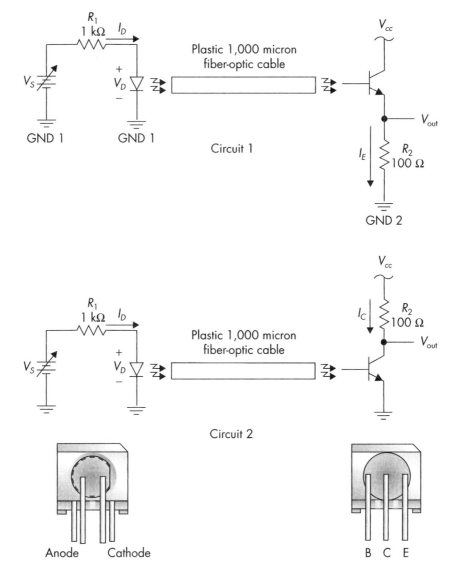

Procedure

NOTE When performing the procedure steps, you have the options of using a calculator or using an Excel spreadsheet program "worksheet" for any required calculations. You may also use Excel for creating tables and for generating graphs.

1. Cut the fiber-optic cable into two pieces. One section is to be three inches long. Measure the length of the longer cable section.
   - Short cable length _____ inches.
   - Long cable length _____ inches.

   NOTE ▶ Be sure to polish all cable ends using 600-grit emery paper. Do NOT exceed a photoemitter voltage drop of 3 volts.

2. **Short Cable Section** Using the short cable section inserted between the photoemitter and the photodetector, connect the circuit shown in Circuit 1.

   V_{CC} is +5 V_{dc}. V_S is a separate supply initially set at 0 V_{dc}. GND1 does not need to be connected to GND2.

   A. With $V_S = 0$ V_{dc}
      1. Measurements. Record answers in Table 1 on page 482.
         a. Measure and record the photoemitter voltage drop (V_D).
         b. Measure and record the photodetector emitter voltage (V_{out}).
      2. Calculations. Record answers in Table 1.
         a. Calculate and record the photoemitter current (I_D), where ($I_D = V_{R_1}/R_1$.
         b. Calculate and record the photodetector emitter current (I_E), where $I_E = V_{out}/R_2$.
         c. Calculate and record the Current Transfer Ratio (CTR), where CTR = I_E/I_D.
   B. With $V_S = 1$ V_{dc}
      1. Measurements. Record answers in Table 1.
         a. Measure and record the photoemitter voltage drop (V_D).
         b. Measure and record the photodetector emitter voltage (V_{out}).
      2. Calculations. Record answers in Table 1.
         a. Calculate and record the photoemitter current (I_D), where $I_D = V_{R_1}/R_1$.
         b. Calculate and record the photodetector emitter current (I_E), where $I_E = V_{out}/R_2$.
         c. Calculate and record the Current Transfer Ratio (CTR), where CTR = I_E/I_D.
   C. With $V_S = 2, 4, 6, 8, 10, 12, 14, 16, 18,$ and 20 V_{dc}.
      1. Measurements. Record answers in Table 1.
         a. Measure and record the photoemitter voltage drop (V_D).
         b. Measure and record the photodetector emitter voltage (V_{out}).
      2. Calculations. Record answers in Table 1.
         a. Calculate and record the photoemitter current (I_D), where $I_D = V_{R_1}/R_1$.

b. Calculate and record the photodetector emitter current (I_E), where $I_E = V_{out}/R_2$.

c. Calculate and record the Current Transfer Ratio (CTR), where CTR $= I_E/I_D$.

3. **Long Cable Section** Record answers in Table 2 on page 483.

   Replace the short fiber-optic cable section with the long fiber-optic cable section. Repeat steps A, B, and C as described in Procedure 1.

4. **Long Cable Section** Using the long fiber-optic cable section inserted between the photoemitter and the photodetector, connect the circuit shown in the schematic (Circuit 2).

   V_{CC} is +5 V_{dc}. V_S is a separate supply initially set at 0 V_{dc}. It is not necessary for GND1 to be connected to GND2.

   A. With $V_S = 0\ V_{dc}$
   1. Measurements. Record answers in Table 3 on page 484.
       a. Measure and record the photoemitter voltage drop (V_D).
       b. Measure and record the photodetector emitter voltage (V_{out}).
   2. Calculations. Record answers in Table 3.
       a. Calculate and record the photoemitter current (I_D), where $I_D = V_{R_1}/R_1$.
       b. Calculate and record the photodetector collector current (I_C), where $I_C = (V_S = V_{out})/R_2$.
       c. Calculate and record the Current Transfer Ratio (CTR), where CTR $= I_C/I_D$.

   B. With $V_S = 1\ V_{dc}$
   1. Measurements. Record answers in Table 3.
       a. Measure and record the photoemitter voltage drop (V_D).
       b. Measure and record the photodetector emitter voltage (V_{out}).
   2. Calculations. Record answers in Table 3.
       a. Calculate and record the photoemitter current (I_D), where $I_D = V_{R_1}/R_1$.
       b. Calculate and record the photodetector collector current (I_C), where $I_C = (V_S - V_{out})/R_2$.
       c. Calculate and record the Current Transfer Ratio (CTR), where CTR $= I_C/I_D$.

   C. With $V_S = 2, 4, 6, 8, 10, 12, 14, 16, 18,$ and $20\ V_{dc}$
   1. Measurements. Record answers in Table 3.
       a. Measure and record the photoemitter voltage drop (V_D).
       b. Measure and record the photodetector emitter voltage (V_{out}).
   2. Calculations. Record answers in Table 3.
       a. Calculate and record the photoemitter current (I_D), where $I_D = V_{R_1}/R_1$.
       b. Calculate and record the photodetector collector current (I_C), where $I_C = (V_S - V_{out})/R_2$.
       c. Calculate and record the Current Transfer Ratio (CTR), where CTR $= I_C/I_D$.

5. Using the data from Tables 1 and 2, graph the photodetector emitter current (I_E) versus the photoemitter current (I_D).

6. After completing the table and producing the required graphs, answer the Analysis Questions and create the brief Technical Lab Report to complete the project.

Table 1 Short Cable Data (Circuit 1)

V_S (V)	V_D (V)	I_D (A)	V_{out} (V)	I_E (A)	CTR

Table 2 Long Cable Data (Circuit 1)

V_S (V)	V_D (V)	I_D (A)	V_{out} (V)	I_E (A)	CTR

Table 3 Long Cable Data (Circuit 2)

V_S (V)	V_D (V)	I_D (A)	V_{out} (V)	I_C (A)	CTR

Analysis Questions

NOTE Answers to these Analysis Questions should be clearly numbered and documented on separate sheets of paper with your name and the date at the top of each page. These answer sheets are to be turned in with the rest of the project documentation, as appropriate.

1. Compare the results seen from the step 2 and step 3 data.
   a. Is V_{out} step 2 the same as V_{out} step 3? (*Yes/No*) Explain your answer.
   b. Is I_E step 2 less than, greater than, or equal to I_E step 3? Explain your answer.

c. Is I_D step 2 less than, greater than, or equal to I_D step 3? Explain your answer.

d. Is step 2 CTR less than, greater than, or equal to step 3 CTR? Explain your answer.

2. Compare the step 3 and step 4 data.
    a. Is V_{out} step 3 the same as V_{out} step 4? (*Yes/No*) Explain your answer.
    b. Is I_E step 3 less than, greater than, or equal to I_E step 4? Explain your answer.
    c. Is I_D step 3 less than, greater than, or equal to I_D step 4? Explain your answer.
    d. Is step 3 CTR less than, greater than, or equal to step 4 CTR? Explain your answer.

Technical Lab Report

Write a brief technical lab report summarizing the technical facts learned from this project. The report should be organized to provide the following:

1. An introductory paragraph describing the type of circuit being analyzed and the key parameters that will be discussed relating to this circuit.

2. A section describing the most important characteristics of this type of circuit that were shown via the collected data in the tables and graphs.

3. Any special facts or characteristics about this type of circuit that were highlighted in answering the Analysis Questions.

4. A practical example of how the information learned in this project might help you in operating, troubleshooting, error analysis, or adjusting a circuit of this type in your home setting, in your training program setting, or in a job setting in the real world.

5. A summary statement listing the most positive aspects of the project and any parts of the project that were difficult because of equipment problems or unclear instructions. Include areas that might be improved.

APPENDIX A
Parts Suppliers Listing

NOTE:

Many parts suppliers will ship small quantities to individuals and/or to school personnel; however, some require minimum orders of from $15 to $25.

Some resources that the author has found useful from time to time are as follows:

RESOURCE NAME	PHONE NO.	FAX NO.	INTERNET
Newark Electronics	1-800-2-Newark	1-800-718-1998	www.newark.com
Digi-Key	1-800-344-4539	1-218-681-3380	www.digikey.com
RadioShack	—	—	www.radioshack.com

Another good resource for finding parts suppliers is the listing found in the back "Reference" section of *The ARRL Handbook for Radio Amateurs* (often simply known as the Radio Amateur's Handbook).

For an exhaustive listing of companies that produce and/or sell parts, you might want to get a copy of *Harris EITD Electronic Industry Telephone Directory*. Contact numbers are: phone: 1-800-888-5900; fax: 216-425-4328.

SPECIAL NOTE:

Possible resources for the new 100-mH (100,000-μH) inductor value introduced in some projects in this 5th edition are as follows:

RADIOSHACK.COM	DIGIKEY.COM
Catalog # 900-4940 (RadioShack)	Part # M 7104-ND (Digi-Key)
Manufacturer's # 70F101AF (J.W. Miller)	Manufacturer's # 70F101AF (J.W. Miller)
Value: 100,000 μF	Value: 100,000 μH
R dc: 278 Ω	I dc max: 29 mA
I dc max: 29 mA	

APPENDIX B

How to Create a Graph or Chart Using Excel

One of the more powerful tools the spreadsheet provides is the ability to easily develop graphs and charts from information you have placed on a worksheet.

It is impossible in this small appendix to give you all the details and ramifications of creating the many types of graphs and charts feasible with the program. We will limit our information to an example of a very basic graph, based on a small collection of data on a sample worksheet.

To do this, we will show you the hypothetical worksheet, and then list the steps you would use to create a bar chart to represent the important data on that sheet.

Try Creating a Graph

Look at the sample partial worksheet shown in Figure 1.

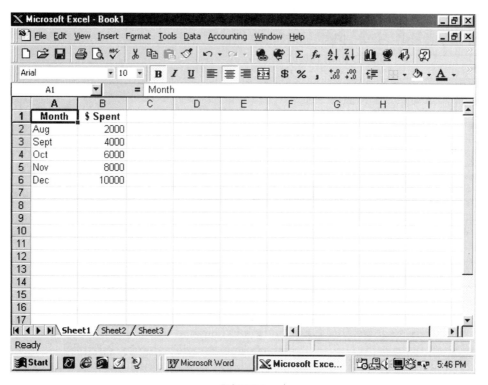

FIGURE 1

To create a graphic display of the expenditures, perform the following steps as we discuss and display each step.

STEP 1 Type the data into a worksheet as we have shown it in Figure 1.

STEP 2 Move the Cell Pointer over Cell B2. (This will indicate that Cell B2 is the active cell.)

STEP 3 Hold down the Shift key and use the down arrow to highlight cells B2 through B6, then release the Shift key. Cells B2 through B6 should be in a big box, with Cells B3 through B6 having a dark (or black) background, as shown in Figure 2.

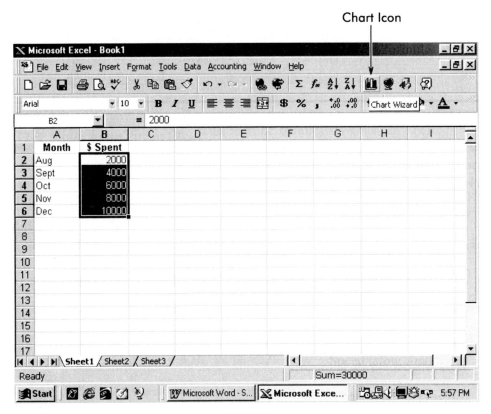

FIGURE 2

STEP 4 Using the mouse, point at and left-click on the Chart Icon near the top right side of the spreadsheet screen, as pointed out by the arrow.

STEP 5 Follow instructions on the "Chart Wizard" subscreens, as appropriate.

Notice in the "Step 1 of 4—Chart Type" screen, shown in Figure 3, that the type of graph selected by default is "Column." Since this is the type we want in our sample, you don't have to point at and click on any of the other types listed. (It is already selected for you.)

NOTE ➤ You can select any of the other types on the list whenever appropriate for your desired purposes.

STEP 6 Point at and left-click on the "Next" button at the bottom of the Chart Wizard screen. "Chart Wizard—Step 2 of 4" is now displayed on the screen, as shown in Figure 4.

You will note in Figure 4 that a sample of the bar graph chart is shown and that the program has identified the specific range of cells you highlighted earlier as the "Data Range" you wanted to use to provide the value data for the vertical component of the graph.

STEP 7 Point at and left-click on the "Next" button at the bottom of the Chart Wizard screen. You will see a new screen, which allows you to fill in some titles

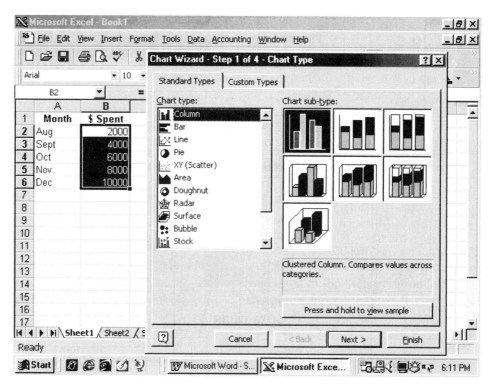

FIGURE 3

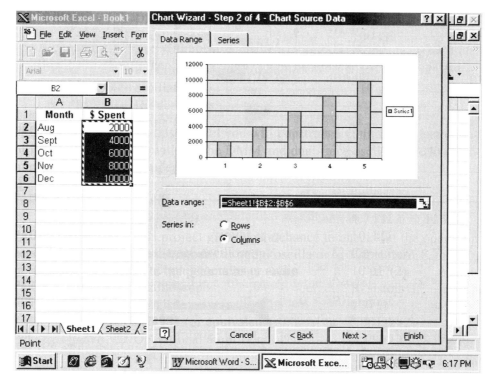

FIGURE 4

or other data you want in order to label the graph and its coordinates. See the screen shown in Figure 5 on page 492.

Observe that this is "Step 3 of 4" in the Chart Wizard sequence. Note that this screen gives you opportunity to type in a "Title" for the graph and to label

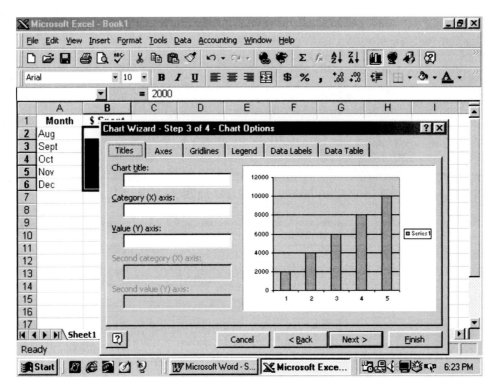

FIGURE 5

the horizontal axis and vertical axis with descriptors, as you desire. For our example, shown in Figure 6, follow the bulleted instructions.

- With the "blinking cursor" in the "Chart Title" box, type in the words "Practice Graph."
- To quickly move the blinking cursor to the next box you wish to type in, press the TAB key.
- With the blinking cursor in the "Category (X) Axis" box, type the word "Time."
- Again, to move the blinking cursor to the next box to type in, press the TAB key.
- With the blinking cursor in the "Value (Y) Axis" box, type the word "Amount."

NOTE ▶ Use the mouse to point at the desired location or box you wish to type in. Slide the mouse until the arrow changes to an "I-beam" pointer in the box you want to type in, then left-click. A blinking vertical bar cursor will appear in that box. When you type, the first letter will appear at the point where the blinking vertical cursor is located!

STEP 8 Point at and left-click on the "Next" button at the bottom of the Chart Wizard. An option screen will appear, which allows you to make the graph "As new sheet" or "As object in" your existing worksheet. Point at and left-click on "As new sheet." Then, click on the "Finish" button at the lower right corner of the Chart Wizard screen.

You should see something like the graph shown in Figure 7.

Voila! You have created a graph! Once you go through these steps a few times, the Chart Wizard will seem self-explanatory.

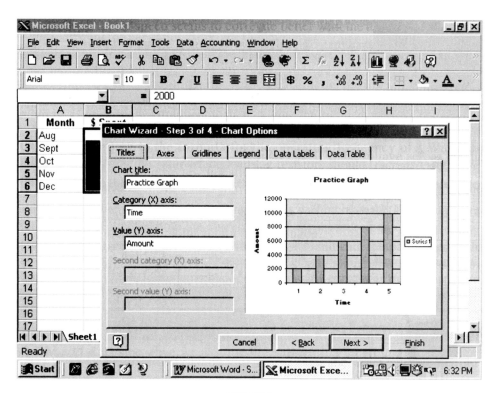

FIGURE 6

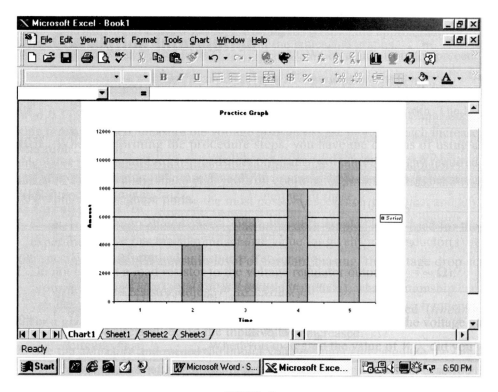

FIGURE 7

With further practice and experience, you will learn more of the bells and whistles that can be added to your graphs. Many things relating to legends, grid lines, data tables, data callouts, and so forth are things that will take you beyond the basics we discussed here. In addition, you can choose different kinds of graphs to illustrate your data. For example, you can choose to use line graphs, pie charts, and more. It is highly probable that, if you try out these variations, you will grow to like and use the powerful graphing tool of your spreadsheet program throughout your technical career.